Smart Moves: Choosing Best Routes

Namita

Contents

Chapter 1

Research background

In transportation planning, decision-making involves evaluating various solution alternatives and choosing the one that best satisfies a policy maker's objective. For the Decision Maker (DM) to get an acceptable solution, the alternatives need to be evaluated against one objective or multiple objectives. Thereafter, their scores would be compared to obtain the *best* solution. Problems of this nature are known as optimisation problems. They are known as Single-objective Optimisation Problems (SOPs) when a DM has only one objective, such as finding the least expensive mode of transport in terms of out of pocket expenses. However, when the problem has more than one objective like choosing the transport mode that is the least expensive in terms of travel fare and environmentally friendliness, such a problem would be characterised as a Multi-objective Optimisation Problem (MOP).

Generally, optimisation techniques are used to identify the *best* among alternative solutions. However, the quality of results the optimisation scheme can find is limited by how the solutions are evaluated. The difficulty of evaluating a solution depends on the nature of the same. For instance, if the solution is a simple phenomenon, that may be described by a linear mathematical equation, evaluating it will be uncomplicated. However, if the solution is a large stochastic system with elements of randomness, as is the case with many real-life problems, the evaluation would be remarkably daunting. In many cases, representing such systems with analytical expressions and attempting to find their closed-form solution is futile. One reason for this is that expressions, variables and parameters cannot adequately describe a system's randomness. A good instance of such a system is a transit network where passengers make *arbitrary* choices in response to the dynamics on the network with the overall aim of maximising their utility.

The focus of this **book** is, therefore, on the design of public transport networks in a way that adequately takes into account this stochastic passenger behaviour. The discipline has been around for as long as there was a need for efficient transportation systems. Indeed most travel-related activities and interactions occur on a pre-designed network. Public Transit Network Design (PTND) involves finding the best network solution, that simultaneously optimises the stated objective(s) of different identified transport stakeholders. Factors to consider in designing them include: 1) how to evaluate or model passenger behaviour, 2) how to represent travel demand data, and 3) the type of optimisation algorithm that is adopted to solve the problem. Therefore, to find a suitable solution, these issues need to be handled appropriately.

Historically, PTND experts have used tools like aggregated Trip-Based Models (TBMs) for travel demand modelling, data from Origin Destination (OD) surveys or OD matrices to represent travel demand and mathematical models to optimise the problem. Yet, if more advanced ones like disaggregated Activity-based Travel Demand Model (ABTDM) replace these tools, will they facilitate the design of networks that better respond to the changing travel demand landscape within the public transport sector? This question among others motivate this research. In the remainder of the work, the terms *public transport* and *transit* will be used interchangeably. Both words refer to scheduled transportation services, with a public

offering that runs along planned routes. The services convey people who are willing to pay, in large numbers, despite having different origins and destinations (Walker, 2011). The next section gives an overview of the current state of the public transport network in the City of Cape Town (CoCT). The city is a typical South African city, and the context of the work reported in this **book**.

1.1 Public transport context of Cape Town

Cape Town is a metropolitan municipality in the Western Cape province of South Africa (see Figure 1.1). It has an estimated land area of 2455 km² (TDA, 2015).

Figure 1.1: Map showing the Western Cape Province and the City of Cape Town, adapted from Viljoen and Joubert (2019).

In the city, public transport planning is done by the Transit Development Authority (TDA), with travel demand modelling and network design being some of the authority's most important activities. According to Cape Town's integrated public transport network plan (RHDHV, 2014), the city's population was 4.04 million inhabitants in 2017, and it is projected to grow to about 4.5 million by 2032. Furthermore, an estimated 68% of the population is of working age. This indicates that a large portion of the city's population is economically active, which translates to very high levels of demand for travel. In terms of the modal network characteristics for the morning and evening peak periods, there is a 53 : 38 split between private cars and public transit among all travellers (RHDHV, 2014; StatsSA, 2017). This modal split excludes non-motorised modes of transport. A closer look at the data by income group shows that 82% of high-income earners use private cars, while, 77% of low and medium-income earners either walk or use public transport. The high dependence on private cars give rise to significant network congestion in the city, especially during the peak commuting periods. Additionally, in terms of other network indicators such as travel time by mode, public transit work trips on average take between 45 minutes by Bus Rapid Transit (BRT) and 63 minutes by bus. Table 1.1 presents further details of the modal share and the network operating indicators by mode for public transport in Cape Town.

Table 1.1: Average travel characteristics of formal public transport modes in Cape Town (TDA, 2014, 2018).

Mode	Modal share (%)	Average route distance (km)	Average speed (km/hr)	Travel time (min)
Rail	23	23.4	18	59
Bus	19	18.1	6	63
BRT	15	20.0	2	45
Minibus taxi	19	21.5	12	53

The current situation as revealed by the information in the table is not sustainable due to the long duration of travel on the various modes of transport. This trend will likely worsen as the population of Cape Town grows and travel demand increases. A general lack of investment in the public transport sector in cities across South Africa in the 1990s is responsible for this state of affairs (COGTA, 2016).

Furthermore, there is little or no physical or inter-modal operational coordination between the four highlighted modes of public transport. This is because of different levels of ownership of public transport infrastructure and operations between the metropolitan, provincial and national governments as well as privately-owned companies. The national government of South Africa, for example, owns and manages the rail services and infrastructure through a state-owned agency known as Passenger Rail Agency of South Africa (PRASA). Private organisations such as Golden Arrow Bus Service (GABS) operate bus services through contracts with the provincial government of the Western Cape region, while the MyCiTi BRT service is owned and operated by the metropolitan government in the city (Salazar Ferro et al., 2013).

To this end, these services plan their respective transit routes and operational schedules with minimal consideration for other modes. Additionally, some sections of the network, such as the Central Business District (CBD) experiences high levels of congestion, while other parts are underutilised. The latter is quite visible in the south-eastern and peripheral parts of the city, which experience a short supply of public transit facilities in some areas or an outright lack of them in others (Behrens, 2004). The aforementioned issues, therefore, make it critical to improve public transportation in Cape Town. Consequently, the expectation is that the work discussed in this **book** would contribute to achieving that goal.

The next section, describes Cape Town's BRT service known as the *MyCiTi* service in more detail. The discussion is relevant because this research would be applied to the BRT network.

1.1.1 The MyCiTi BRT

The MyCiTi BRT service began its operations in 2010 as part of a nationwide rollout of BRT services across South Africa. In Cape Town, the service is expected to be a significant component of the Intergrated Public Transport Network (IPTN) which is a public transit network planned in anticipation of the future effect of urban growth on travel demand in the city. The plan involves a significant expansion of the city's current public transportation network. This is logical given the expected growth of the city's population by approximately 11% in 2032 when the IPTN is scheduled to be fully functional. According to the TDA (2014), a vital objective of the IPTN is to ensure that at least 80% of the inhabitants in the CoCT will live within 500m of a BRT trunk or rail line by the target year of 2032. The backbone of the MyCiTi is a full specification BRT network which currently provides transport services along the western axis of Cape Town. The system utilises recent technologies like Automated Fare Collection (AFC), closed transfer facilities and level boarding platforms. It will be rolled out in four stages, with the full system ready for service in about twenty years. The first phase routes of the service were launched officially in 2011. Since then, new routes have been incrementally developed to expand the service's coverage within the city. Currently, two express services are undergoing testing between the Cape Town CBD and the city's South-Eastern axis for the second phase of operations.

1.1.2   The MyCiTi base network

The MyCiTi BRT network functions as a trunk-feeder system. Currently, the network consists of 472 nodes and about 46 operational routes. The feeder lines typically involve circuitous movements as the service mainly collects and distributes passengers from residential areas unto trunk lines and vice versa. Conversely, the trunk routes carry passengers along their travel desire lines while preventing geometric irregularities and trip indirectness as much as possible. Trunk-feeder networks can improve operational efficiency by closely matching public transport demand and supply (Wright and Hook, 2007). The MyCiTi feeder services use smaller vehicles that go through residential areas to give passengers access to terminals and transfer stations, that facilitate their connection to other parts of the network. The principle behind this type of service is that smaller vehicles are less expensive to acquire and operate; hence, they run in low demand areas. Therefore, they are more suitable for providing frequent services in low demand areas. The stop locations have different configurations, namely trunk stations or feeder couplets. Trunk stations are closed areas on the system, that allow passengers to enter the system before they board a vehicle. The station platforms allow travel in either direction along a route. On the other hand, the feeder stop couplets, are open areas consisting of two feeder stops on opposite sides of a roadway. Typically, feeder stops service passengers on different route directions, with the type of stop facility determining the nature of passenger interaction on the MyCiTi system. This work deals with the improvement of the MyCiTi trunk network. Figure 1.2 shows the image of the network.

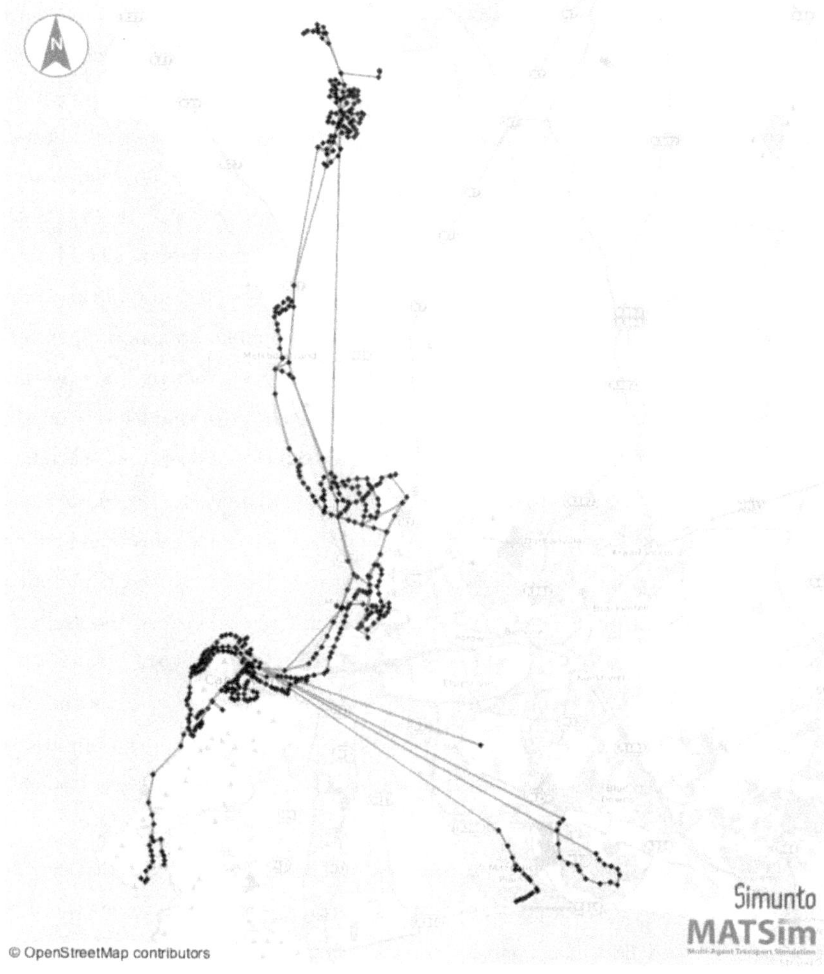

Figure 1.2: Map of the MyCiTi BRT network.

1.1.3 Operational challenges on the MyCiTi service

Currently observed trends in the planning and design of the MyCiTi BRT reveals the need to improve the service on different fronts such as network optimisation, operational cost reduction and even improvement in terms of its administrative structure. Officials at the TDA indicate that current trip patterns are not conducive to operate economically sustainable transit services. The primary travel demand problems recognised by the MyCiTi service planners is that of low patronage on the network. Hence, without a substantial change in the way commuters travel or user perception of the service, the revealed demand on existing routes would not change.

Another impact of the low passenger ridership on the MyCiTi BRT is that it impacts the operators' costs and subsidy requirement of the system. The MyCiTi service, like most formal or scheduled public transport systems around the world, operate on some form of subsidy. Subsidies enable public transport operators to maintain a stated and acceptable level of service even though; they are not economically sustainable. At the current revealed travel demand levels on the MyCiTi service, passenger revenue would not cover the costs of operation. Therefore, subsidies would be needed to fund the resulting operational deficit. Hence, there are benefits to be gained from increasing ridership on the system since this would reduce the amount of subsidies required to operate the service. Lastly, the MyCiTi business plan, as developed in 2012, reveals a significant difference between the system's projected costs and revenues and that of the operational data in 2014 (TDA, 2015). Based on this observed trend, there is a considerable risk of costs exceeding the available subsidy provisions. As a result of this, the city of Cape Town initiated a rationalisation exercise for the MyCiTi service to address growing concerns revolving around financial sustainability (Nnene et al., 2017).

The network-related aim of the rationalisation exercise was to optimise the network routes and to reduce operating costs through effective route design, schedule optimisation and fleet management, among others. The process comprised an in-depth review and adjustment of operational practices and service characteristics throughout the MyCiTi network. It entailed surveying and analysing all routes in detail.

On the whole, the rationalisation exercise gave insights into the key drivers of costs within the system. The process enabled the CoCT to identify the possible trade-offs between cost reduction and quality service delivery on the system. Further details of the rationalisation exercise are available in TDA (2015) which describes other significant cost-saving measures. In line with this, the next section discusses the design of public transport networks as one possible way of improving public transit systems. The concept of transit network design is defined and situated within the broader transportation planning discipline. Hence, the discussion also touches on some difficulties associated with designing public transport networks.

1.1.4 The design of public transport networks

A public transport network is a collection of stops or stations connected in space by a sequence of links, that make it possible to operate transit services. The links can either be tangible, such as rail lines and road lanes or intangible as in the routes connecting a network of airports. Transit networks play a significant role in facilitating the socio-economic activities that occur in any geographical area. In economic terms, they are the primary means of satisfying travel demand or the *supply* side of the transportation planning cycle. Designing public transit networks is a challenging endeavour. A major source of complexity in the design of public transit networks stem from the unique topography of each network based on the context, travel demand and demography of people that interact with it. These factors influence decisions such as 1) type of service to plan for; 2) technology to deploy for the service; 3) operational considerations like policy on speed, frequency, and fare; 4) what section of the population should the service prioritise. Other sources of complexity are a large number of human agents with their arbitrary decision-making and interactions on the network. There are also many non-linear occurrences, like congestion, where a localised incident such as a vehicle crash can lead to network-wide travel delays. Moreover, the linkage between transit networks and other external dynamic systems like the environment and economy creates additional complexities (Rodrigue et al., 2013).

Solving a transit network design problem generally involves finding a network, and its accompanying service frequencies, that best satisfies a stated goal. Over time, researchers have formulated various solution schemes to address some of these issues. These solution techniques are generally classified as either analytical or heuristic. The former can find a unique solution to the Transit Network Design Problem (TNDP) using analytical algorithms. In the literature, analytical solution models are criticised because they are hardly applicable to large scale transit network problems. This is mainly because getting a closed-form expression for the objective function is computationally too expensive, if not impossible (Chakroborty, 2003). On the other hand, heuristics by its very nature, cannot search for an exact solution to the TNDP. They instead obtain suitable approximate solutions of a global optimum solution—assuming the latter exists. In any case, their solutions are typically considered acceptable given the relatively smaller amount of time they use to find the solution. For this reason, heuristics, especially meta-heuristic algorithms, have been widely adopted with a positive outcome in the solution of large scale TNDPs.

In recent times, the latter has been used more often, due to its wide adaptability to different problem domains. The components of a typical TNDP solution framework are alternative transit network solutions, user demand, operational schedules, decision variables, stakeholder objectives, and resources like a vehicle fleet. The next section discusses the specific research problem statement in the next section.

1.2 Problem statement

When solving a TNDP there are three concerns to address. The first is how to model people's travel demand. Another concern is how to represent the problem's decision variables such as network route configuration and service frequency. The final consideration is if the network design and frequency setting sub-components of the problem is modelled sequentially or simultaneously. Consequently, a solution model for the TNDP should fully describe stochastic traveller behaviour, optimise a transit network and its operational frequency simultaneously, and define a suitable decision variable representation. Along these lines, the major research problem in this book deals with developing a suitable method to address some identified limitations associated with existing solutions in the literature. Three of these problems are discussed here as follows:

- There should be a feedback loop between public transport network routes and their service frequencies: This implies that the problems of route design and frequency setting should be tackled simultaneously to satisfy travel demand effectively. However, owing to the considerable difficulty associated with tackling each of these problems, researchers often address them sequentially. This approach is problematic because, in reality, an iterative loop should exist between the demand level on a route network and the available fleet when designing transit networks.

- The inability to define a suitable encoding scheme for the problem variables within the solution framework: As a typical optimisation problem, defining a suitable representation for the variables of the TNDP is a very crucial task. Therefore, it is necessary to encode the problem's decision variables in a way that addresses both network design and frequency setting as one problem. In the literature, many works use simplified but largely inadequate representations. However, these are insufficient to describe the problem's network and frequency decision variables and their operations.

- The inherent limitation of conventional TBM (which has been use in most published TNDP publications) in analysing or assessing public transport system objectives when they deal with the microscopic details of individual travellers. This is because they are aggregate models and are better suited for strategic and regional studies.

A TNDP solution model that would be considered an improvement on existing ones must address the earlier mentioned issues in a better way. The model should be capable of describing the stochastic interactions of travellers on a transit network. It should also have a suitable way to encode or represent decision variables within the optimisation framework. Lastly, the model should be able to facilitate the simultaneous optimisation of a public transit network and its operational parameters.

After developing the proposed method, it will be applied to the design of public transit networks in Cape Town to assess its computational performance and its applicability to real life transit network problem.

1.2.1   Research objective and questions

In light of the preceding discourse, the overarching objective of this research is *to develop a TNDP solution model that uses a Simulation-based Optimisation (SBO) approach.* The proposed solution technique should be capable of addressing the earlier identified gaps. The method should also be useful for designing large scale transit networks. The key research questions in this **book** are therefore

- *How can a SBO solution model for the transit network design problem be developed through integrating Agent-based Travel Demand Simulation (ABTDS) and Multi-objective Optimisation (MOO)?*

- *How can the model be applied in the design of public transit networks in the city of Cape Town, South Africa?*

The next section conceptually describes the proposed SBO transit network design framework.

1.3   Research Design

1.3.1   Simulation-based optimisation

The proposed network design model involves a synthesis of both simulation and optimisation. An optimisation algorithm is generally used to find a unique or acceptable solution to a problem, given a set of alternatives. Simulations, on the other hand, through experimentation, can describe the complex behaviour of an entity within a system. They replicate large-scale, real-life systems as a means of gaining a better understanding of the system's performance under different scenarios. Notably, optimisation and simulation models exhibit complementary limitations in their functionality.

Optimisation algorithms can find solutions but do not fare well in analysing systems—especially large stochastic ones. Conversely, simulations may explain the latter but cannot optimise them. This complementary behaviour allows using both phenomena in achieving the aims of this **book**. The combination of optimisation with simulation to solve decision-based problems such as the TNDP is known in the operations research literature as SBO. The discipline is relatively recent, even though optimisation and simulation have been around for a very long time as separate disciplines. The development of techniques, such as *simultaneous perturbation* for non-linear optimisation (Bhatnagar et al., 2013), and *reinforcement learning* for dynamic programming, have advanced the evolution of optimisation methods that are compatible with simulation models and the modification of older optimisation algorithms to more easily integrate with simulation (Gosavi, 2015a). Furthermore, advancements in computational science have aided the overall growth in this field. As a result, it is now possible to process larger models in much less time. In a typical TNDP, the problem search space comprises of networks with varying configurations and their corresponding operational schedules. Each network solution is evaluated by simulating travel demand on it. At the end of the simulation, the performance indicators are used to rank the networks. This process continues repeatedly until an efficient solution for the problem is found. Hence, the optimisation algorithm ranks and chooses each network solution based on the results of simulating the solution.

In this **book**, the proposed SBO public transport network design combines a MOO algorithm with an Agent-based Simulation (ABS) for use in the network design process. The MOP is used because the TNDP is primarily a problem with more than one objective. Usually, the objectives are conflicting, and the goal is to find a solution that best represents a compromise between the objectives. Conflicting objectives on a transit network are visible in the fact that travellers prefer direct routes which reduce the time spent travelling to a given destination. On the other hand, transport service operators make allowance for slight deviations from direct routes though it is at variance with the users' desire. Such deviations enable the operator to cover more demand, thereby increasing their profit and reducing average operational costs. Hence, in this work, the goal will be to find the most efficient solution that reduces network utilisation costs for passengers and operational costs for operators. ABS is used to evaluate each solution, as mentioned earlier. This simulation technique models the disaggregated activity and behaviour of individual travellers on the network, rather than people's aggregated travel behaviour as done in mainstream travel demand modelling.

1.3.2 Justification for the use of agent-based modelling

The use of ABTDS in this work is based on the advantages of the model over the conventional trip-based transport planning models. Some of these are discussed in Castiglione et al. (2015) including the fact that Agent-based Model (ABM) can assess transportation policies that are difficult to test using convention TBMs. For example, ABMs provide better sensitivities analysis when evaluating costing and pricing scenarios. Also, because ABMs function at the level of individual travellers and represent how the trips are made over the whole day, the model is more sensitive to pricing and other policy objectives that may vary with the time of day. Another important advantage of ABMs over conventional TBM is that they can provide more detailed performance metrics for specific public transport systems and studies. In addition, they also produce all the TBM indicators that may be used to support transport studies involving regional public transport travel demand forecasting. However, despite these advantages, a major shortcoming of ABM is that they are data intensive and very computationally expensive models. This is due to microscopic level of data needed to build the models. However, as public transport systems are increasing planned with a data-driven philosophy, it is important to develop tools and techniques that are better suited to adapt to these changes.

The next section discusses the implementation steps for the proposed network design model.

1.3.3 Simulation-based transit network design

The interaction between MOP and ABS lies in translating the optimisation model's solution (network) into a format that is readable by the simulation. There would also be a need, to reconfigure the simulation's output into a scheme the optimisation can use to rank the network solution. It is this interaction between the optimisation and simulation models that make the proposed solution model a SBO of public transit networks. This is because the optimisation is being achieved through the agency of a simulation. A graphical representation of a conceptual framework for the proposed Simulation-based Transit Network Design Model (SBTNDM) is shown in Figure 1.3. The framework details the steps taken to achieve the goals of this research.

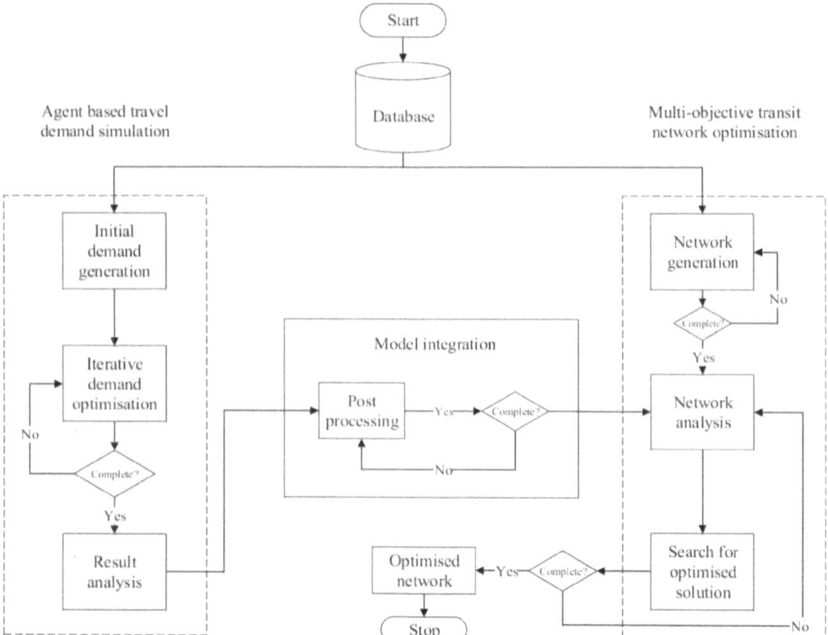

Figure 1.3: Conceptual framework of the SBO framework.

Multi-objective public transit network design

Three main steps are involved in network design, namely, network generation, network evaluation and lastly, a procedure used to search for an optimised network.

Network generation In this stage of the model, computer based heuristic algorithms would be used to create a population of feasible solutions for the network design problem. Feasibility criteria like network size and route length guide the process. This pool of feasible networks is then initialised within the SBO solution algorithm.

Network analysis This involves calling the ABS to evaluate the network solutions initialised in the generation stage. The scores and other indicators obtained from evaluating the solutions, then serve as input for the model's final stage.

Search for optimised solution In this final stage, the integration between simulation and optimisation occurs. The results of the network analysis (simulation) are used alongside other sub-process of the optimisation to rank and compare the solutions. The best solutions are then used to create newer solutions. This process continues iteratively until a predefined termination criterion is satisfied.

Agent-based travel demand simulation

The three main steps of the ABS are as follows:

Initial demand generation In this step, a synthetic population of public transit users are created by sampling from a real population. The individuals in the synthetic population are then assigned demographic and other attributes based on the census data of the transport area under consideration. Subsequently, a daily *plan* comprising all the activity locations and trip chains in the transport area is created. The trip chains are created by organising sequentially the trips embarked on by the travellers within a 24-hour modelling timeframe.

Iterative demand optimisation This stage comprises of three sub-steps. In the first sub-step called *execution*, plans generated by agents in the demand

generation are simulated using an agent-based traffic assignment. The next
sub-step is called *scoring*. Here, the executed plans are evaluated and assigned
a score, using a scoring or utility function. The scores are then used to measure
the performance of a plan and if it should be adjusted or not. The last step,
known as *replanning*, is a process that allows every individual to recalculate
their current plan based on how satisfactory they find the present network
conditions. Replanning facilitates the correct simulation of passenger behaviour,
such as choosing a different route when the initial one they select is congested.

Result analysis At the end of the simulation, various performance indicators
relating to the specific objectives of the study may be collected and analysed to
gain insight into the travel demand and behaviour of agents within the study
area.

1.3.4 Contributions to existing knowledge

To the best of the author's knowledge, a SBO solution framework has never been
used as a technique for solving the TNDP. Therefore, the main scientific contribution
of this work is in *developing a technique that combines simulation and optimisation
techniques to design public transit networks*. The other contribution of the work will
be seen in the *application of this method to the design of a real public transit network*.
The method would address the gaps in the literature as follows:

- To solve the problem of defining a suitable encoding scheme for the problem's
  decision variable, a unique and innovative data structure is created. The latter
  can represent the network and its detailed operating schedule fully assigned
  with the vehicle as one entity.

- In terms of solving the network design and frequency setting problems sequentially
  rather than simultaneously. The earlier mentioned decision variable encoding
  will facilitate the simultaneous optimisation of the transit network and its
  operating schedules.

- The issue of the limited analysis of a public transit network at the microscopic resolution of the network user is addressed will be addressed by simulating passenger network interactions with an ABMs.

1.4   Research methodology

The methodological framework used in this **book** is in line with the *design research* philosophy, which is proposed by Manson (2006). This approach to research involves the utilisation of existing knowledge to develop a model or artefact that will, in turn, be tested and studied rigorously or applied to an existing problem to generate new knowledge. The five main components of the methodology are adapted for this **book** and presented in Figure 1.4.

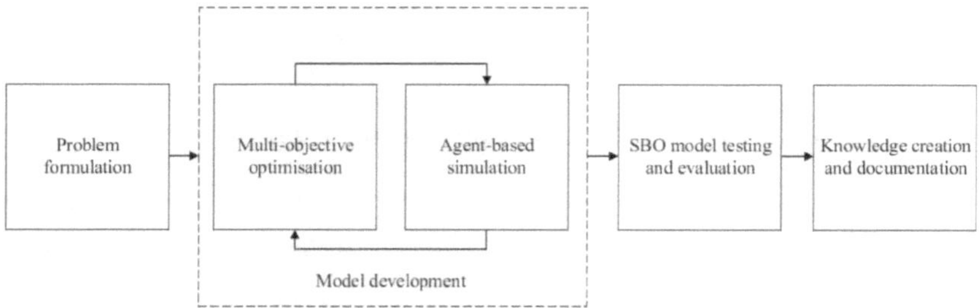

Figure 1.4: Steps of the SBO model.

1.5   book structure

After giving an overview of the proposed research, the structure of the remainder of this **book** is shown in Figure 1.5.

Chapter 2 presents a review of the relevant literature. The primary goal of the review is to describe the key concepts that are relevant to the **book**. It also aims to situate this research within the broader public transport planning and network design literature. Next, in chapters 3 through 5, the discourse focuses on the development and implementation of the proposed model.

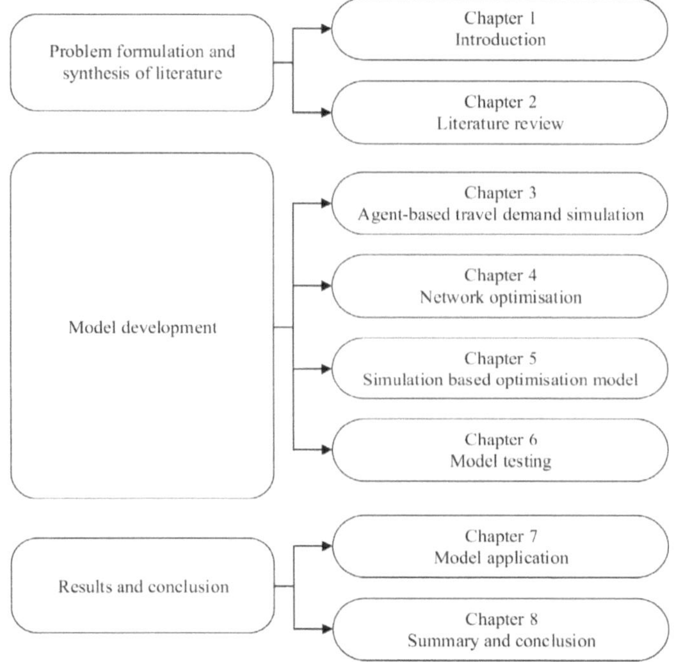

Figure 1.5: Overview of the book.

The three chapters respectively discuss ABTDS, Multi-modal Public Transit Network Design (MPTND) and the integration of both to realise a SBO transit network design model. Chapter 6 focuses primarily on testing the computational performance of the model. In Chapter 7, numerical tests are conducted to evaluate the model. The tests involve applying the proposed SBTNDM to a large-scale TNDP case study. Lastly, in chapter 8, a summary of the work is written and conclusions drawn on this basis. The chapter reflects on the critical points in the book, and the lessons learnt. It also indicates potential directions for future research.

1.6 Publications

Some research articles published by the author and his supervisors while writing this book are listed below.

- Nnene, O.A., Joubert, J.W. and Zuidgeest, M.H., 2019. An agent-based evaluation of transit network design. Procedia Computer Science, 151, pp.757-762.

- Nnene, O.A., Joubert, J.W. and Zuidgeest, M.H., 2019. Transit network design with meta-heuristic algorithms and agent-based simulation. IFAC-PapersOnLine, 52(3), pp.13-18.

- Nnene, O.A., Zuidgeest, M.H.P. and Beukes, E.A., 2017. Application of metaheuristic algorithms to the improvement of the MyCiTi BRT network in Cape Town. Journal of the South African Institution of Civil Engineering, 59(4), pp.56-63.

Chapter 2

Literature review

The introductory chapter presented a background for this book. It highlighted the motivation for the work. The chapter also stated its primary goal as the development of a Simulation-based Optimisation (SBO) model, which harnesses the strengths of optimisation and simulation in the design of public transit networks. This chapter presents a literature review. The objective is to create a foundational understanding of the main concepts used in this work. It does so by synthesising knowledge from the review of various relevant aspects of the literature. The first section discusses models and simulation in the context of public transportation systems. This discussion leads up to a description of Agent-based Simulation (ABS), which is the specific type of simulation used in the book. After discussing literature on simulations, the next section addresses the state of the art in the optimised design of transit networks under the themes of Transit Network Design Problems (TNDPs) and Multi-objective Optimisation Problems (MOPs) respectively. Finally, in the last section, the SBO literature is reviewed.

2.1 Travel demand models and simulation

Transportation planners and engineers often use scientific knowledge and technology to develop various public transport solutions that facilitate safe and affordable mass mobility (Arentze and Timmermans, 2000, 2004). A comprehensive knowledge of human travel behaviour within the dimensions of space and time would increase the likelihood of developing such high-quality transport systems. However, gaining a complete understanding of people's travel needs and behavioural patterns as they go about their activities within a defined geographical space is almost impossible. Therefore, tools that help experts and policymakers alike improve their understanding of this crucial input into the transport system planning process are required.

Models are such tools, as they facilitate the conceptual framing of large natural systems either physically or in abstraction (Lee and Vuchic, 2005). Usually, modelling is done at scales that are small enough for the interested party (planners, researchers and engineers) to study a system and understand the detailed behaviour of the variables in the system (Ortuzar and Willumsen, 2011). Models, therefore, come in handy when a direct investigation of a natural system is either too large, expensive, unsafe or disruptive (El Sheikh et al., 2007; White and Ingalls, 2016). Travel demand models afford us a systematic way to investigate how people's demand for public transport services change as a result of various input assumptions (Castiglione et al., 2015). Such models replicate the behavioural interactions of agents within the transportation system.

The first travel demand models were built in the United States of America in the 1950s. The discipline grew out of the need to correctly predict future travel volumes, which served as an essential design parameter in the development of highway systems being constructed at the time (Bates, 2000). Since then, two dominant modelling philosophies have evolved in transportation planning, namely *trip-based* and *activity-based* modelling. Trip-Based Models (TBMs) are so named because the trip is the basic unit of analysis. They represent travel demand in terms of aggregated inter-zonal trips. Hence, they are also known as *aggregated* travel demand models and are more suited for regional and sub-regional public transport analysis. The Activity-based Model (ABM) use people's activities as the dominant unit of analysis. Therefore trips are modelled as the connection between activities. ABMs are *disaggregated* models because they are implemented with microsimulation frameworks, using data that reflect the travellers' choice and decision-making in microscopic details. Other classifications for travel demand models are *static* and *dynamic*. This deals with how the models distribute or assign travel demand on a network in the temporal dimension. TBMs are static and are only able to capture average performance indicators for a specific time snapshot on a transit network. On the other hand, ABMs are dynamic as they can capture time-dependent phenomena on the network, particularly at the level of passenger activities like boarding or alighting a transit vehicle (Duell et al., 2015). The transportation model that is of particular interest to this **book**, is the ABM particularly a class of these models known as ABS. The latter models travel demand through simulations.

In general, a simulation is an experiment-based method of studying and analysing the behaviour of a model's variables and parameters (Perros, 2009; White and Ingalls, 2016). It is used to gain a better understanding of a natural system or phenomenon. When the simulation is in the form of a computational model, it is called a simulation model. In public transport planning, simulations help us analyse the detailed interactions of users, drivers and other stakeholders. They give a better understanding of how these entities interact within the transportation system under different prescribed scenarios. The lessons learnt may then inform decision-making when providing transit solutions. The insights and information received can also be used during network upgrades and other activities that help to improve the attractiveness of a public transportation system.

Simulations have been used to study pedestrian movement in public transit infrastructure (Bohari et al., 2016), bus priority system planning (Papageorgiou et al., 2007), bus passenger behaviour at stops (Cortés et al., 2011) and passenger response to service disruptions (Yin et al., 2016). Different types of simulations in the literature include discrete event, continuous system, agent-based, Monte Carlo, and hybrid simulations. More details regarding simulations are available in Schruben (2017); White and Ingalls (2016). The type of simulation utilised in this research is known as ABS. The next section discusses it in the context of public transit planning.

2.1.1   Agent-based simulation

Understanding the concept of *agents* is central to grasping how ABSs work. An agent is an autonomous entity that possesses some attributes and exhibits specific behaviour within a modelled environment. They are *autonomous* because they receive information from their environment and change their state or behaviour in response. Furthermore, agents do not require human intervention to act. They operate on their thread of execution within the model, and they interact with other agents and with their environment. With this in mind, a model that replicates a real-life system, using an abstraction built around agents is called an *agent-based* model. Such a model must specify the rules that govern the attributes, behaviours and environment of the agent. Hence, agent-based modelling aims to search for analytical insight into the collective behaviour of agents that obey simple rules, especially in natural systems (Uhrmacher and Weyns, 2009). This feature distinguishes Agent-based Models (ABMs) from multi-agent systems where the goal in the latter is to solve specific computational or scientific problems through the action of a group of agents.

According to Heath and Hill (2010), the theoretical foundations of ABS were laid by researchers like Hebb (1952) who attempted to understand the occurrence of emergent behaviour in systems. This was done by studying individual agents and using them as the catalysts for the characteristic patterns observed in the system. Emergent behaviour refers to actions and behaviours that are observed among the agents when they react and adapt to system changes within their environment (Bonabeau, 2002). Usually, such behaviour is not part of the original rules that define the agents' actions. Siebers et al. (2010) asserts that this *emergent phenomenon* results from the input variables of the simulation and the agent's interaction with other agents. Emergent behaviour, which is at the heart of ABS, was first observed in the cellular automata game, *game of life*, developed by John Conway (Weimer et al., 2016). In the game, checker-like objects replicate themselves and form patterns on a grid, to observe the way complex patterns emerge from the execution of straightforward rules.

Over the years, with the advent of computers, ABS as a discipline has grown from such knowledge areas like cellular automata (Chopard et al., 2009; Russell and Norvig, 2016) and heuristics (Albar and Jetter, 2009). Some notable differences between agent-based simulations and other types of simulation are:

- agent-based simulation is more complex than others like discrete event simulations, as the former demonstrates emergent behaviour.

- structurally, the agents are set up in an object-oriented programming framework, where each agent is a separate entity with its memories and rules;

- the abstraction of agent-based simulation models is different from others because the philosophy revolves around the agent, their goals and decisions;

- they provide a more detailed and natural description of a system;

- each agent has its thread of execution or control within the simulation framework.

In this section, the concept of ABS is discussed. The next section would look at how the ABS evolved as a formal research interest in academia, particularly in the context of transportation.

2.1.2 Agent-based travel demand simulation

In transportation modelling, Agent-based Travel Demand Simulations (ABTDSs) are viewed as a sub-class of Activity-based Travel Demand Models (ABTDMs) (Castiglione et al., 2015). The pioneering study on ABTDMs is found in Mitchell and Rapkin (1954). The authors established the crucial link between people's activities and the demand for travel. They also made a case for having a more accurate way of modelling travel behaviour. Furthermore, the work shows that ABMs were conceptualised around the same time as TBMs. However, the latter enjoyed greater acceptance due to its perceived ability to better respond to the transportation policy directions of that era, which mostly leaned towards the provision of highway infrastructure (McNally and Rindt, 2008). Nonetheless, over time, transportation planning problems have become more sophisticated, requiring more advanced planning tools and techniques. Today it is no longer sufficient to only forecast traffic volumes for highway construction, but to also manage the demand for the infrastructure and its utilisation after its implementation.

By the 1970s, there was a renewed interest in modelling an *individual*'s trip in connection with their activities. Examples of such works are that of Chapin (1971);

Cullen and Godson (1975). Hägerstraand (1970) is also credited for research on the distribution of opportunities in the space-time dimensions, and how it imposes some constraints on people's ability to participate in activities. He conceptualised three major constraints that influenced travel behaviour, as follows;

Authoritative constraints which are enforced by institutional authorities: the number of working hours fixed by the management of an organisation, or transportation schedules issued by public transport agencies.

Capability constraints which refers to natural and material restrictions: biological needs such as the need to go home and rest after a workday.

Coupling constraints which point to restrictions that arise from the need to interact with others when performing some activities: waiting for other team members before starting a departmental meeting in an office.

This study formed the basis of the space-time prism concept discussed in Bhat and Koppelman (1999); Pinjari and Bhat (2011), and the time-geography theory by Vuk (2001). Chapin (1971) extended the ABM discourse by proposing that a combination of societal constraints and an individual's motivations were the driving forces behind an individual's activity pattern. Cullen and Godson (1975) undertook empirical studies to show the degree of rigidity or flexibility between Hägerstraand's space-time constraints. Their results showed that the time constraint was less flexible than the spatial limitation: the nature of the activity a person engages determined the flexibility of the time constraint. For instance, a person would tend to be less flexible with their resumption time at work on a weekday than that of recreational activities in the weekend.

By 1983, the Transport Studies Unit (TSU) at Oxford University developed the first empirical ABTDM (Jones et al., 1983). The author synthesised the earlier discussed concepts to develop the model. Despite these advances, the modelling technique largely remained in the academic domain because of the widespread acceptance of TBM by industry practitioners. However, since the early 2000s activity-based modelling has gained more approval in the industry. Hence ABMs have been applied to several cases like the ABS of transport demand and land use (Huynh et al., 2015), mixed land use and transport (Ziemke et al., 2016), and urban freight modelling (Bok and Kwon, 2016). Other applications are in traffic behaviour modelling (Hager et al., 2015), parking choice behaviour modelling (Habib et al., 2012), and bundling of transportation into urban agent-based models (Wise et al., 2017). More applications are discussed in Horni et al. (2016).

In this work, ABTDS is used as a component of the proposed Simulation-based Transit Network Design Model (SBTNDM). In an ABTDS, the agents include network users, operators, vehicles and other entities. One assumption is that an individual's decision to travel depends on their need to engage in an activity (travel demand is a derived need). The activity depends on the agent or decision-maker, a set of activity options or *choice sets*, and specific heuristic rules that govern the sets and define the boundaries outside which they become unrealistic. The major components of the model are *activity-based demand generation* and *agent-based traffic assignment*. These are discussed, respectively, in the next two sections.

Activity-based demand generation

In general, Activity-based Demand Generation (ABDG) involves the following:

Generate agent population In this step, a *synthetic population* of agents is generated, using a concept of *random realisation* described in Balmer et al. (2006). This involves creating a virtual population that has the same demographic attributes and structure as that of a real population census. Essentially, if one takes from the synthetic population, within statistical limits, it should return the real census. Additionally, the synthetic population is composed of spatially located households, which have specified attributes such as a street address, car ownership and household income. The households are then assigned individuals having attributes like *gender* and *age*.

Generate individual activity schedules This step consists of generating *activity schedules* for each respective agent in the synthetic population. Activity here is defined simply as a *physical engagement of an individual in something that satisfies his personal and/or family needs* (Vuk, 2001). A daily schedule reflects the pattern of activities to be engaged in by the agent as well as information like start and end times of activities, duration of activity and the activity location.

Passenger mode choice In this last step, mode choices are made by the agents for each trip. This is similar to the third step in the four-step model (mode choice), but it does not only distribute the trips based on modal characteristics. Instead, modes are also selected based on the demographic attributes of the travellers. An instance is that only families with car ownership as an attribute will have the option of selecting that mode for their trips.

Normally, transportation planning models have *demand* and *supply* components. The demand side focuses on people, their attributes (socio-economic and demographic) and their travel demand. Conversely, the supply side deals with the transit network, its utilisation per user demand and the assignment of the demand. These concepts are also applicable to ABTDM. The discussion in this section focused on the demand side—travel demand generation, in the context of agent-based models. This involves generating a synthetic population of travellers, creating daily activity plans for the population and defining their mode choices for the trips that will connect their activities. The next section explores the supply component, namely agent-based traffic assignment.

Agent-based traffic assignment

Nagel and Flötteröd (2009) formalised the already existing concept of agent-based traffic assignment. It was further developed as a part of the implementation of an ABTDM known as Multi-Agent Transport Simulation (MATSim) (Nagel and Flötteröd, 2016). According to the authors, the research was motivated by the need to extend the *behavioural dimensions* in existing Dynamic Traffic Assignment (DTA) models, and to account for more *realistic* and *dynamic* constraints under which travellers make their decisions. To meet their research goals, the authors logically extend the definitions of static and dynamic traffic assignment to cover the perspective of the individual traveller. Hence, a full description of the agent-based traffic assignment will benefit from a brief description of the static and dynamic traffic assignment models.

In a transit network, people choose their travel route based on the cost associated with the network's routes. Hence, the user behaviour witnessed on a network largely depends on the attendant cost of travelling on that network. This behaviour, which is reflected in the traveller's route choice, in turn, changes the network's state relative to congestion and delays. The new conditions on the network, subsequently, affects the route choices of the traveller, which again changes the network's state. When this cycle progresses to a point where the route choice of the traveller can no longer affect the network as to improve their travel experience, a point of User Equilibrium (UE) has been attained (Wardrop, 1952). One assumption used in Wardrop's UE is that the travellers have a perfect and uniform valuation of all route costs at all times. Hence, they will always choose the *least expensive* route (de Dios Ortúzar and Willumsen, 2011).

However, this is not realistic, owing to the stochastic and sometimes irrational nature of human decision-making. It implies that people will not always choose the shortest travel path. Therefore, the UE model has been extended to account for the randomness associated with people's decision-making (Sheffi, 1986). The principle guiding this modification is that each user *perceives* travel costs differently, and their perception guides their choice of least expensive route(s). Their choices, therefore, leads to a distribution of routes, with the most attractive route having the highest demand. This extended model is known as the Stochastic User Equilibrium (SUE) assignment model. The critical limitation of the SUE is that, behaviourally, the model only accounts for a traveller's route choice. This limitation makes them insufficient to fully describe the microscopic behaviour of agents on transit networks.

Furthermore, the Wardrop's and stochastic UE models consider static traffic assignment and are typically represented with a structure as (origin, destination). This is because they do not account for time-dependent variations of travel demand and travel costs on a transit network. To this end, Friedrich et al. (2000); Kaufman et al. (1991) propose some improvements to the static assignment that account for temporal variations. The enhanced model is known as DTA. It incorporates an additional dimension for *departure time* in its structure: (origin, departure time, destination). Its major contribution is the ability to generate time-dependent traffic or link volumes. Therefore, the agent-based traffic assignment model is a direct extension of the DTA (Chiu et al., 2011).

The keystone of agent-based traffic assignment is in decoupling the steady flow of passengers on the network in the static and dynamic contexts to that of the individual traveller. Flötteröd and Rohde (2011); Zhou and Taylor (2014) show that it is challenging to model traffic flow dynamics in complex networks. However, disaggregating the Origin Destination (OD) matrix into individual trip makers allows for vehicle assignment to each trip maker. Additionally, considering individual travellers rather than a steady flow of passengers is computationally less burdensome (Zhou and Taylor, 2014).

Based on this premise, the UE and SUE can then be extended to a so-called *disaggregate* or *particle* case, where the particle represents the *microscopic* or single traveller. This allows the model to retain the structure of the DTA—(origin, departure time, destination)—while permitting the linkage of individuals' activities into an activity chain or plan. One consideration is that the plans must be linked in such a way that certain logical constraints are respected. This is done so that the model is a realistic description of travel behaviour. For instance, a subsequent trip should start where the former one ended.

Lastly, similar to the case of SUE, each agent has a choice set of activity plans from which they choose the plan to execute. The stochastic nature of the agent's decision, therefore, leads to a choice distribution model. Resolving this would be too computationally expensive as it would entail analysing every plan in each agent's set of plans. Hence, in Balmer et al. (2006), a Genetic Algorithm (GA) is used to create new plans for the agents through its crossover and mutation operators. This allows the evolution of better plans for agents by analysing and scoring their selected plans, with the consideration that each choice set of plans is a population of plans in the genetic algorithm. Also, since this operation coincides for the agents in competition with other agents for network resources, the model is set up as a co-evolutionary algorithm framework (Meneghini et al., 2016).

The discussion on ABS completes one part of the proposed SBTNDM. The next section deals with the design of transit networks using Multi-objective Optimisation (MOO) techniques. It starts with a general introduction of the TNDP as a sub-discipline within the broader public transportation planning. The ABTDS process can be seen in Figure 2.1.

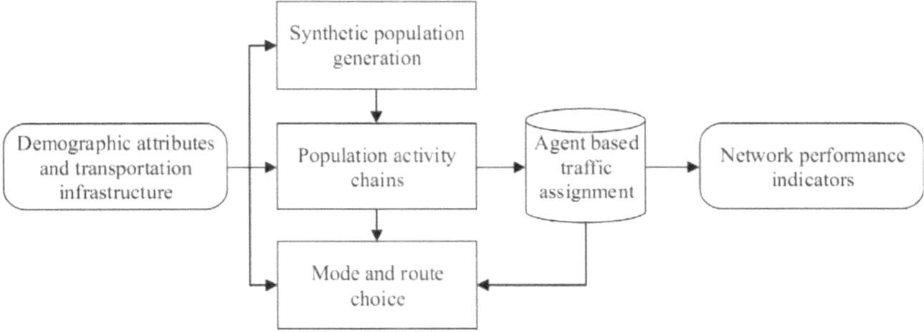

Figure 2.1: Flow chart for agent-based travel demand simulation, adapted from Zheng et al. (2013).

2.2 Public transit network design and optimisation

Public transportation planning is an expansive subject area that incorporates many interrelated activities that are carried out by experts to deliver quality transit services. Vuchic (2005) splits the discipline into three sub-disciplines. They include *public transit systems operations and networks, public transit economics, agency and organisation*, and *systems planning and mode selection*. The three give a broad understanding of the various skill sets required in the public transport sector such as engineering, planning and economics. More challenging is aligning these skills and interests in an interdisciplinary fashion, to serve the purpose of quality service delivery.

The first sub-discipline focuses on the transit network and its operational components. They involve route network modelling and optimisation, as well as service scheduling. The second sub-topic in Vuchic (2005) deals with the spectrum of management, organisation and economic issues that occur in the provision of transportation services. They include system financing, transit fares, transit agency organisation and regulations, subsidies, and transit system ownership models. The final aspect considers and evaluates different system components. It addresses the issue of appropriate technologies, suitable transit modes, and how they interact to achieve the goal of serving a community with an attractive, affordable and sustainable transportation service. Ceder (2015) further decomposes the network planning and operational process into a four-step procedure, namely network route design, timetable development, vehicle scheduling and crew scheduling. The first two listed items form the basis of discussions in the next section of this work. The public transport planning process is illustrated in Figure 2.2.

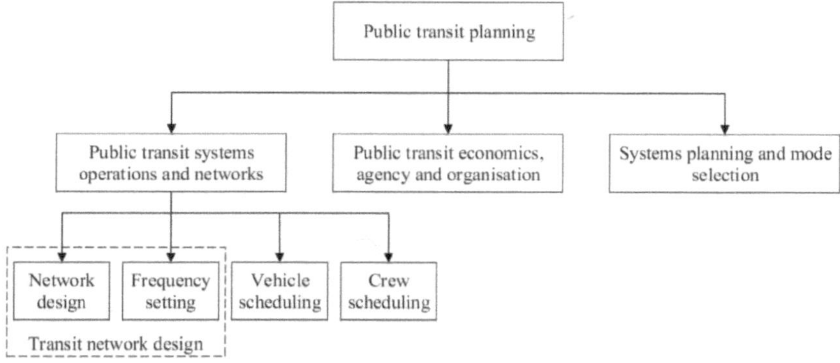

Figure 2.2: Components of the public transport planning process, adapted from Avishai (2007).

2.2.1 Transit Network Design Problem (TNDP)

Transit network design is a complex and rigorous endeavour. It involves finding a set of routes and their operational frequencies that best address the stated goal(s) of multiple network stakeholders. It is an optimisation problem, subject to a set of discrete or continuous constraints such as network configuration, route choice and service headway. Other considerations are the feasibility constraints on route

length, vehicle capacity as well as fleet size. According to Ceder (2015); Ibarra-Rojas
et al. (2015), the network design and frequency setting activities of the TNDP are
respectively classified as strategic and tactical, while vehicle and crew scheduling are
operational activities. The network design is seen as strategic because it involves
long term planning and requires less detail. Conversely, frequency setting is done in a
shorter time horizon but involves a higher level of detail. Hence, its description as a
tactical activity. Lastly, the vehicle and crew scheduling or operational activities take
place in the shortest time and requires microscopic route level details (see Figure 2.3).

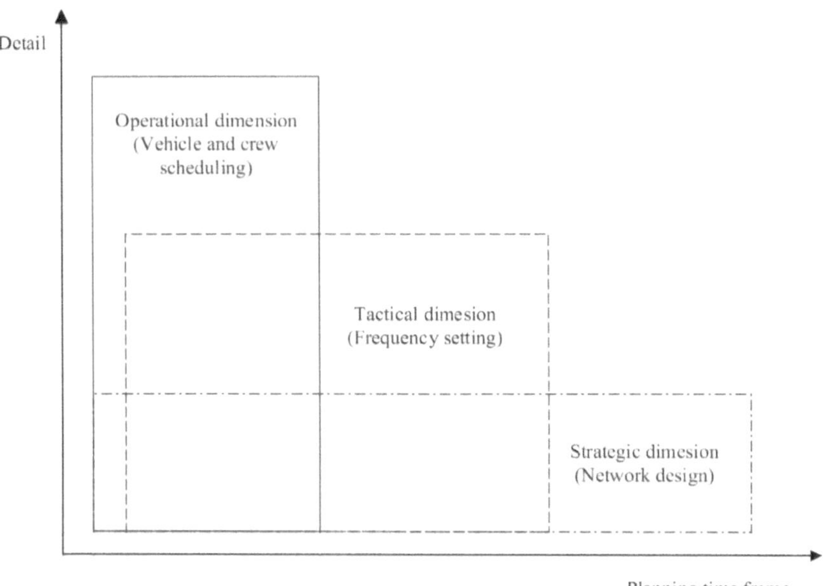

Figure 2.3: Transportation planning time horizons, adapted from Zuidgeest (2005).

The above description highlights the features of the TNDP as a combinatorial
multi-objective optimisation problem. It is *combinatorial* in the sense that the goal
is to find an optimal set of routes and their operating frequencies among a finite set
of alternatives (Schrijver, 2003). It is also a *multi-objective* optimisation problem
because of the many conflicting objectives of different stakeholders such as network
users, operators and even the public transit authorities (Buba and Lee, 2018; Talbi,

2009c).

Solution methods for the TNDP are categorised as analytical and heuristic. Both solution techniques are modelled mathematically. The primary distinction is the way they search for a solution. Analytical methods utilise *exact* algorithms such as the *branch and bound, branch and cut* or the A^* algorithm. They attempt to find the closed form of the objective function in the search for a unique solution to the problem. Instances of research done with analytical solutions in the literature are Chen et al. (2017); Chien et al. (2001); Daganzo (2010); Ouyang et al. (2014). Conversely, heuristic techniques use *approximate* algorithms which can find *good* solution(s) in a reasonable amount of time. Some examples of such research works are Cipriani et al. (2012); Nikolić and Teodorović (2014); Szeto and Wu (2011).

The major criticism of analytical methods is that by nature, the TNDP is a *non-convex* (Baaj and Mahmassani, 1995; Newell, 1979; Shih and Mahmassani, 1994) and *np-hard* (Chen et al., 2017; Fan and Machemehl, 2006, 2012) problem. This implies that their equivalent decision problem is *np-complete*. Hence there is no verifiable efficient algorithm that exists to solve them in polynomial time (Fan and Machemehl, 2004; Talbi, 2009c). Therefore, it is almost impossible to derive a closed-form expression for its objective function. The other criticism of analytical solution methods is their limited scope of application to idealistic or small-sized problems. This implies that applying them to more realistic TNDPs would require an almost infinite amount of computational resources, which are unrealistic. More recently, the earlier mentioned criticism has led to the development of a type of heuristic algorithms known as *meta-heuristics*. Their main advantage over the general heuristic is that they are not problem dependent (Fan and Machemehl, 2004). Hence, they are amenable to a broader array of problems spanning various knowledge domains than ordinary heuristics. Dréo et al. (2006) highlight the fact that when the number of objectives in a MOP exceeds two, meta-heuristics are best suited for the problem. In terms of the TNDP, meta-heuristic algorithms can be applied to large scale or realistic problems since they can find acceptable network solutions within a limited period. Meta-heuristics would be adopted as the solution technique of choice in this book.

Lastly, as a typical optimisation problem, the main components of the TNDP are: objective function(s), decision variable, feasibility constraints and a user behaviour sub-model. These elements are discussed briefly in the next section:

2.2.2   Components of the TNDP

Objective Function An objective function expresses the intended *goal* of an optimisation problem. Typically, it is expressed mathematically. Depending on the nature of the optimisation problem, a single objective or multiple objectives may be used in the formulation of an objective function. They are typically presented as the minimisation of a cost factor or the maximisation of a benefit or utility. In many real-life MOPs, the objectives are generally contradictory; hence solving an optimisation problem is an attempt to find the best compromise solution between conflicting objectives. van Nes and Bovy (2000) discusses six different formulations of the objective function which have been used in the TNDP. They are, namely the minimisation of total travel time, and total transit cost. Others are the maximisation of the following: transit operator profits, cost-effectiveness and total satisfied demand.

Decision Variables In the TNDP literature, a decision variable is a resource which the policymaker wants to optimise (Curtin, 2004). A feasibility condition usually defines the limits or bounds of their availability. Some of the most commonly used decision variables in the design of transit networks are route alignment, route frequency, route length, headway, service frequency and timetables (Ngamchai and Lovell, 2003; Shih and Mahmassani, 1994).

Feasibility Constraints These are parameters that reflect the limiting conditions of the decision variable in a transit network design problem (TNDP). They generally define the feasibility of the optimisation problem and ensure one obtains solutions within reasonable resource limitations. Some of the commonly used constraints in the literature are maximum/minimum service frequencies, maximum load factor, route length, fleet size and operational cost. Pattnaik et al. (1998) used a combination of maximum frequencies, allowable fleet size and maximum load factor. Cipriani et al. (2012) used route length and service frequency to ensure that the length of the resulting routes and their service frequencies does not exceed an acceptable threshold. These constraints ensure that network solutions are generated within realistic limits, defined by the availability of resources or their scarcity thereof.

Travel demand sub-models The travel demand sub-models in the TNDP usually serve as a proxy for passenger or user behaviour. They help to describe people's trip making choices and decisions within a transit network. The four-step travel demand model has been used extensively for this purpose in the literature, especially the traffic assignment step of the model. While static transit assignment sub-models like the All-or-nothing (AoN) and UE have mainly been used in the literature, agent-based traffic assignment (Horni et al., 2016; Nagel and Flötteröd, 2009), will be utilised in this **book**. The method facilitates the disaggregated assignment of individual traveller's trips on a network. Hence, it supports the SBO network design approach that is proposed in this work.

2.2.3   Review of the TNDP literature

The review here focuses on solution models that have previously been used by researchers to solve the problem. Some notable review articles on the subject area are those by Buba and Lee (2018); Fan and Mumford (2008); Fan and Machemehl (2004); Farahani et al. (2013); Ibarra-Rojas et al. (2015); Johar et al. (2016); Kepaptsoglou and Karlaftis (2009). In these papers, the authors give a detailed classification for the TNDP. Three prominent categories in the articles are optimisation objective(s), solution techniques and parameters like *decision variables, demand pattern, network structure*, among others. Besides, the earlier mentioned review articles, the section also describes the works of other notable authors in the literature that are relevant to this **book**. The reviewed works span the five decades, since, the first TNDP research was documented. They are grouped into two main publication periods, which occur before and after the year 2000.

Pre 2000

The work of Lampkin and Saalmans (1967) is the first documented research in the
TNDP literature. The central theme of the work was to redesign an existing bus
network in a city in England, for improved efficiency. Since the loss of patronage/traffic
to private cars was the primary motivation for the work, the specific optimisation
objectives were to maximise the overall network utilisation and minimise travel time
subject to fleet size availability. They defined route spacing and frequency as the
problem's decision variables. For their travel demand sub-model, the authors used
the generation, distribution and assignment steps of a typical four-step model. Lastly,
the authors used a heuristic solution model based on a greedy search algorithm to
find the most efficient network of routes.

In the three decades between when Lampkin and Saalmans's work was published,
and the year 2000, most of the documented research works in the literature have
implemented heuristic and analytical solution techniques or a combination of both.
Ordinary heuristics have mostly being used to generate feasible network or route
alternatives, while the optimisation problem is solved with an analytical model. A
deviation from this trend is the work of Pattnaik et al. (1998), who implemented a
GA based meta-heuristic solution.

Dubois et al. (1979) proposed a hybrid two-stage model, which comprised both
route network design and frequency setting. The route configuration and service
frequency parameters were used as the decision variables, while the optimisation
objective was to minimise the generalised time of travel subject to constraints
on investment cost. The solution involved choosing a set of routes and their
corresponding bus lines with a modified version of Lampkin and Saalmans (1967)'s
link insertion technique. In the second step of the model, transit line frequencies
were calculated using an analytical model. The author assumed a variable demand
context to get the optimal frequency.

LeBlanc (1988) also proposed an analytical transit network solution model, which
had an operating frequency as its only decision variable. The author's optimisation
objective was to minimise the operator's cost and maximise transit usage with
constraints on operating cost. A simultaneous modal split - traffic assignment model,

was used to consider the frequency of every transit line because the changes in modal split on some transit lines affected the service frequency of others. The solution model was used to design a public transit network in North Dallas, Texas.

Finally, in Constantin and Florian (1995), the authors proposed an analytical solution approach based on a sub-gradient algorithm. The objective of their model is to minimise waiting time and total travel time, while, satisfying fleet size constraints. Route frequency served as the decision variable in their model, while a multipath traffic assignment model was used to assign travel demand on the network. This solution approach was used to improve a sample transit network in the United States.

Lastly, the details of other significant works published in the period under review includes those proposed by Baaj and Mahmassani (1995, 1990, 1991); Ceder and Israeli (1998); Chakroborty et al. (1998); Clarens and Hurdle (1975); Higgins et al. (1996); LeBlanc and Abdulaal (1984); Magnanti and Wong (1984); Mandl (1980).

Post 2000

Since the 2000s, meta-heuristic algorithms have been used more often as solution models for the TNDP. This indicates that in more recent times, meta-heuristic solutions have gained more acceptance. One reason for their widespread acceptance is their relatively simple adaptation to optimisation problems even when large and extensive case studies are to be solved. Also, the advances in the fields of operations research and computational science in the last two decades have made the implementation and use of meta-heuristic procedures relatively straight forward.

Some notable works that have used meta-heuristic solutions from the year 2000 till date include An and Lo (2015); Arbex and da Cunha (2015); Chakroborty (2003); Chen et al. (2017); Cipriani et al. (2012); Fan and Machemehl (2004, 2006, 2008); Huang et al. (2018); Lee and Vuchic (2005); Madadi et al. (2019, 2018); Nnene et al. (2019a).

However, all the mentioned research works adopt single objective optimisation algorithms in their solutions, even though the TNDP is essentially a multi-objective problem. One of the major differences between single and multi-objective solution algorithms is that in the former, a linear summation is used to reduce many objectives to a single one. Furthermore, weights have to be defined beforehand for each objective. The obtained single results then reflect the weighted objectives. On the other hand, the outcome of multi-objective algorithms is a set of solutions representing all the possible trade-offs between a problem's objectives. This makes it possible to obtain valuable information about the trade-offs between the objectives and sensitivity for weighting the various objectives in terms of an optimal design solution (Possel et al., 2018). However, this, among other reasons, make the multi-objective solution approach more complex that the single-objective. To use single-objective algorithms, researchers often make simplifying assumptions at the risk of rendering the obtained network solution idealistic or hardly applicable to realistic large scale networks. Understandably, the assumptions are made to reduce the complexity and intractability of the TNDP.

In the literature, when a TNDPs is solved with multi-objective solution models, it is called a Multi-objective Transit Network Design Problem (MTNDP). Fewer of these works exist in the literature when compared to those solved with single objective optimisation procedures. Next, some of these works will be discussed since they are pertinent to this book. Also, most authors now use multi-objective meta-heuristic algorithms as their preferred solution in their research.

In Brands and van Berkum (2014) the authors proposed a bi-level programming framework as their network design solution model. Their optimisation objective was to minimise the total (generalised) travel time, the number of car trips to and from urban zones, CO_2 emissions and public transport operating deficit. The decision variables defined in the work includes the location of park and ride facilities, train stations and the frequency of public transport lines. A SUE transit assignment model was used to assign travel demand within the solution framework, and it represents the lower level objective that deals with traveller behaviour on the network. A multi-objective GA known as the Non-dominated Sorting Genetic Algorithm (NSGA-II) was used to as the solution algorithm. The final network results were obtained by doing multiple runs of the model and taking a mean of the obtained results to account for the randomness in the model. The model was tested on a multi-modal transport network between the cities of Amsterdam and Haarlem in the Netherlands.

In their work, Possel et al. (2018) describes a multi-objective transport network design problem. The problem was presented in a bi-level framework. The upper level represents the transport planner task, which is to determine traffic measures that minimise the objective functions in terms of travel demand. Next, the lower level explained the travellers' behaviour that optimises their objectives. For traffic assignment, the authors used a UE model. The solution model was applied to the Almelo transport network in the eastern part of the Netherlands.

Sharma et al. (2009) proposed a MTNDP with the objective(s) of minimising vehicular emissions and total system travel time under budgetary constraints. This was also framed as a bi-level problem, with the upper-level problem dealing with the transport planner's desire to obtain optimal road capacity-expansion vectors while minimising the travel time and emission objectives. The lower level problem dealt with user behaviour on the network and a deterministic UE assignment model represented it. They solved the problem with a NSGA-II. The model was used to solve for a real large-sized network in the city of Pune, India.

Finally, Heyken Soares et al. (2019) developed a model to solve the Multi-modal Public Transit Network Design (MPTND). The objectives were to minimise the travel time for passengers and the total operating distance for the operator. They used a heuristic traffic assignment model, which involves, 1) identifying the terminal nodes within a network, 2) creating a node-to-node demand matrix based on a catchment area created around each terminal node and, 3) assigning half the total demand on each terminal on the link connected to the terminal. Their solution model was an implementation of the NSGA-II. The model was used to solve for a real large-sized network in the city of Nottingham, United Kingdom. In the next section, multi-objective optimisation will be elaborated on.

2.2.4 Multi-objective optimisation

The TNDP has been characterised as a typical MOP. Problems of this nature have at least two conflicting objectives that need to be resolved simultaneously. Therefore, solving them requires making certain *trade-off* decisions to obtain an *efficient* or *compromise* solution that is acceptable (Elarbi et al., 2017). Many real life optimisation problems, fall into this category. A general mathematical statement for MOPs is presented in (2.1)

$$
\begin{cases}
\min : f(x) = [f_1(x), f_2(x), f_3(x), \ldots, f_N(x)]^T & \\
g_j(x) \geq 0 & \forall j = 1, \ldots, J \\
h_k(x) \geq 0 & \forall k = 1, \ldots, K \\
x_i{}^{Lower} \leq x_i \leq x_i{}^{Upper} &
\end{cases}
\tag{2.1}
$$

where N is the number of objective functions ($N \geq 2$), J and K are the number of inequality and equality constraints respectively. $x_i{}^{Lower}$ and $x_i{}^{Upper}$ symbolise the lower and upper bounds of the decision variable. If a solution x_i satisfies the $J + K$ constraints, it is said to be *feasible*. The set of all feasible solutions constitute the feasible search space. The case above is framed as a minimisation problem. However, it can be converted to a maximisation by multiplying each objective function by -1 and transforming the constraints based on the *principles of duality*, highlighted in Deb (2001). The result of an MOP is a set of trade-off solutions known as *Pareto optimal* or *non-dominated* solution(s). Finally, the image of the latter in the objective space is known as the *Pareto front* (Knowles et al., 2008). The concept of Pareto optimality and dominance is discussed in the next section.

2.2.5 Dominance and Pareto optimality

The result of a Single-objective Optimisation Problem (SOP) is usually a single solution. In that case, the quality of the solution can only be measured relative to an optimal solution, assuming the latter exists or is known. In contrast, the result of MOPs is a set of solutions. In most cases, it is hard to identify any single solution in the set that performs better than the others out-rightly on all the objectives. This situation gives rise to the idea of *domination*, which is a way of comparing MOP solutions by looking at how they perform on the various objectives of a problem. In the literature, a solution x is said to dominate another solution y, if x performs better than y in at least one objective, and is not worse than y on all other objectives. If both conditions do not hold, then solution x does not dominate y. The figure 2.4, depicts the feasible search space of a MOP with two objectives (objective function one is to be maximised while objective function two is to be minimised). It gives a clearer picture, of the definition, and aids in the description of the concept of *dominance* and *Pareto optimality*.

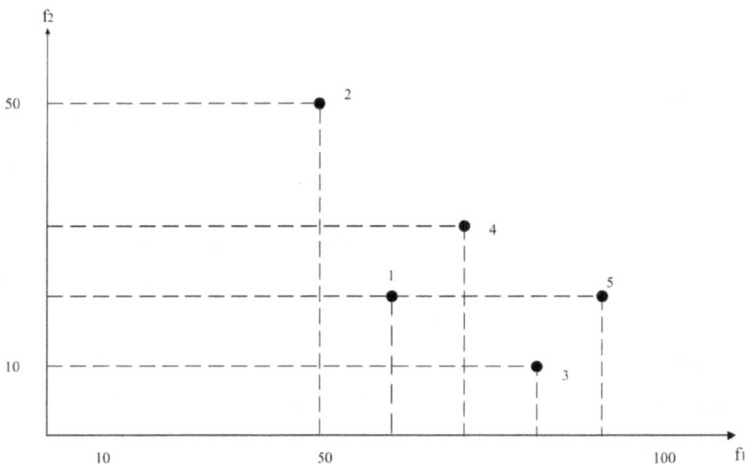

Figure 2.4: Illustrating the concept of dominance, adapted from (Deb, 2001).

In the figure if solution one and two are to be compared, based on the problem objectives and the definition of domination, it is clear that *solution one* is better than *two* on both objectives, hence, *solution one* dominates *two*. Additionally, if we compare solutions *one* and *five*, it is evident that *solution five* performs better on objective one, whereas both solutions perform equally on the second objective. Therefore, based on the conditions of domination, it may be said, *solution five* dominates solution one.

However, comparing *solutions three* and *five*, reveals that while *solution five* is better than *solution three* on objective one, it is indeed worse than *solution three* in terms of the second objective. Therefore, it cannot be said, that *solution five* dominates *solution three* or vice versa. In cases like this, the solutions are considered as *non-dominated* relative to each other. Hence, for any given problem, the set of solutions that are non-dominated relative to each other is called the *non-dominated set*. The properties of non-dominated sets, are; 1) no two solutions within the set dominate any other solution 2) the solutions in the set dominate all other solutions outside the set that belongs to the feasible search space of the problem. The non-dominated set is said to exhibit *Pareto optimality* and it is known as a Pareto optimal set. It is also called a globally Pareto optimal set if it exists and constitutes of all the non-dominated solutions in the feasible search space. The next section focuses on how to get a Pareto optimal set of non-dominated solutions.

2.2.6   Identifying non-dominated solutions

Finding non-dominated solution sets plays a crucial role in this **book**. The concept is needed to obtain an optimised network solutions. According to Deb (2001), the process of identifying a non-dominated solution set is similar to finding the least or greater number in a set of real numbers. However, in this case, the objective is to find a better solution among the lot. Three different computational methods have been used to identify non-dominated solutions; they are as discussed next.

Naive and slow this is essentially a brute force approach, that iteratively compares each solution with all other solutions with a search space. If a solution that dominates the current one is found, the latter will be marked as being dominated and left out of the non-dominated set. However, if any solution does not dominate it, it is inserted in the non-dominated set.

Continuously updated method is similar to the previous method. One solution is added to an empty set $S = \emptyset$ from the population. The next solution, chosen from the population, is then compared with the one in S. If the solution in S is dominated, it is replaced by the one that dominates, and if both solutions are non-dominated relative to each other, the new one is added to S. However, if the new one is dominated by the former, it is not added to the set. In this way, each unique solution taken from the population is compared with the non-dominated solutions contained in S.

Kung et al. (1975)'s efficient method sorts all solutions in descending order of priority, relative to the first objective function of the optimisation problem. A bisection of the population into two sub-populations — top (T) and bottom

(B) follows this. Usually, the *top half* performs better than the *bottom half* relative to the first objective. A check for domination is done between solutions in B and those in T. Subsequently, the non-dominated solutions in B are merged with those in T, to form a *merged* population.

In the literature, Kung et al.'s method is more widely used than the others, because it is seen as being computationally more efficient. The algorithmic details of the three approaches can be found in Appendix A. The next section looks at how MOPs are solved.

2.2.7 Solving multi-objective optimisation problems

Solving a MOP involves finding a set of solutions that is the best approximation of a Pareto optimal set within, the problem's feasible search space (Elarbi et al., 2017). To achieve this, two factors must be considered, namely *convergence* and *diversity* (Talbi, 2009d). Convergence is about the proximity of a solution set to the Pareto set. Indeed, it is a mandatory requirement if desirable solutions to the problem must be found. Convergence is also applicable to SOPs, where the closest solution to a global optimum is sought.

On the other hand, diversity deals with maximising the *distribution* of solutions in a search space. The quality of trade-off solutions is indicated by how much they are spread out over the entire Pareto region. A wider spread across the front indicates that an intensive exploration of the search space has been effected. This implies that the solutions in a non-dominated set should cover all regions of the Pareto front and replicate the curvature of the underlying front as accurately as possible (Deb, 2001; Talbi, 2009e).

Lastly, the diversity within a Pareto set points to the fact that the non-dominated solutions have not been obtaining from a cluster of solutions within a local neighbourhood in the search space. It is also one of the factors, that distinguishes MOPs and SOPs, as it does not apply to the latter.

The complexity of a MOP determines the type of scheme that will be used to solve it. This complexity typically arises from elements, such as the number of objectives, size of the problem's variable space, and the number of solutions in the Pareto front. Others are the structure of the Pareto front in terms of continuity, convexity and multi-modality (Chong and Żak, 2013). Additionally, the nature of a problem's design variables, parameters and constraints, also contributes to its complexity. For instance, as more design variables are added to a problem, its dimensionality increases, hence, its complexity (Michalewicz and Fogel, 2000).

Multi-objective optimisation solution algorithms are basically classified as *exact* and *approximate*. Exact or analytical optimisation algorithms are more effective for solving problems with small search spaces. However, Dréo et al. (2006) highlights the fact that when the number of objectives in a MOP exceeds two, approximate solution methods are best suited for the problem. This is because of the *hardness* of such problems and their multiple-objective frameworks.

However, they don't guarantee that a Pareto optimal solution set would be found. Instead, they ensure that a good approximation of the Pareto optimal solution can be obtained in a reasonable time frame. Heuristic procedures are typical approximate optimisation algorithm. Recall that meta-heuristic algorithms are the solution algorithm of choice in this work.

A class of meta-heuristics that is particularly amenable to multi-objective problems is the multi-objective meta-heuristics. An instance of this is the Multi-objective Evolutionary Algorithm (MOEA). The latter is a type of meta-heuristics that are designed, based on biological principles (Branke et al., 2008; Elarbi et al., 2017; Rangaiah and Bonilla-Petriciolet, 2013). They have been extensively researched, because, of their ability to generate good approximations of the Pareto sets in many multi-objective optimisation problem scenarios and disciplines. The next section explores the MOEA in detail.

2.2.8 Multi-objective Evolutionary Algorithm (MOEA)

MOEAs are multi-objective meta-heuristic search procedures that can be used to find efficient solutions to optimisation problems. Their operations mimic biological phenomena like genetics, bee or ant colonies; hence, they are called *bio-inspired* algorithms. The class of algorithm work, by enabling the realisation of newer and presumably better generations of solutions from existing ones. A typical MOEA framework consists of a population of solutions or chromosomes. Each chromosome is made up of genes that depend on the particular representation of the chromosome. Furthermore, the algorithms have operators whose action on the current population, gives rise to offspring solutions. The latter is generally assumed to be fitter or perform better than their *progenitors*. The next section discusses the design of MOEAs. This focuses on the issues that affect their ability to find suitable solutions to multi-objective optimisation problems. Subsequently, the evaluation of MOEAs is explored.

2.2.9 Design of MOEAs

The factors to consider when designing a MOEA are *fitness assignment*, which facilitates a good convergence of the solution towards the Pareto optimal set; *diversity preservation*, which ensures that diversity is maintained among the generated non-dominated solutions; *elitism*, which guarantees the performance improvement of the algorithm by preserving *elite* (the best) solutions; These factors are discussed below.

Fitness assignment The process of determining the quality or *fitness* of different
individuals, within a population of solutions is called *fitness assignment*. In
single-objective evolutionary algorithms, the operation maps fitness values to
each solution. The values are obtained from evaluating the solutions against
the objective function. In contrast, MOEA algorithms have multiple fitness
assignment strategies, namely scalarisation, criterion-based, dominance-based
and indicator-based methods. The *dominance based* or *non-dominated ranking*
approach are the most relevant to this **book** as it is used in the network optimisation
stage of the proposed SBO model. They mainly identify different sets of solutions
within a population and rank them based on their dominance. The main feature of
this process is that unlike a method such as scalarisation, they do not require
transforming the MOP to a SOP. This is because they are capable of generating
Pareto sets from multi-objective search

spaces. Furthermore, they guarantee better diversity in the Pareto optimal set (Zitzler et al., 2002). Lastly, dominance-based fitness assignment strategies can evaluate each solution across multiple objectives relative to the whole population of solutions (Zitzler et al., 2004). Three common examples of the non-dominated ranking of Pareto solutions are; *dominance rank, dominance count* and *dominance depth*. The dominance rank method involves ranking a given solution, based on the number of solution(s) that dominates it. This strategy, was first used in the Multi-objective Genetic Algorithm (MOGA) by Fonseca and Fleming (1993); Van Veldhuizen and Lamont (2000). Their work represents, an extension of the pioneering work of Goldberg (1989), where the author first applied dominance measures to genetic algorithms. Conversely, the dominance count method, ranks solutions based on the number of solutions they dominate. This method can also be used together with other dominance-based factors. Lastly, in the dominance depth method, the population of solutions is organised into multiple fronts in order from the non-dominated solution to the dominated ones. The non-dominated solutions are assigned a rank of 1 and form the initial front say F_1. Next, the solutions that are only dominated by F_1 are assigned a rank of 2, and they form the second front F_2. The depth of a solution then matches the rank of the front to which it belongs.

Diversity preservation Diversity preservation in MOEAs ensures that search space is explored more broadly and efficiently (Segredo et al., 2017). The main reason for this loss of diversity referred to as *genetic drift* is because of the use of finite population sizes (Talbi, 2009d). It also occurs, due to the biased sampling that sometimes happens during the search for feasible solutions. Hence, the diversity preservation strategy works by worsening highly dense clusters of similar solutions, located in specific local neighbourhoods of the search space. This way, the bias of selecting those solutions is reduced (Zitzler et al., 2004). The need for diversity preservation, is because of the tendency of meta-heuristic algorithms, to experience premature convergence. This issue arises, when members of the population are in a suboptimal part of the search space, from where they cannot escape.

The most suitable methods, for preserving diversity in MOEA solution sets, are; diversity-based approaches such as *Kernel, nearest neighbour* and *histogram* methods (Talbi, 2009d).

Kernel methods, specify the diversity of a solution based on a Kernel function $\gamma(\sigma_d)$ that estimates the density of a solution based on its distance from other solutions in the population. The function γ takes the distance d between solutions σ as its argument. To estimate the Kernel density of a given solution, the distance between it and all solutions in the search space is estimated. Next, the Kernel function is applied to all the distances. Then, the density estimate of the solution is represented by the sum of the Kernel functions. In the histograms approach, the search space is partitioned into several grids, which delineate each neighbourhood in the search space. The density of a solution is the number of solutions in the same box as that solution. In terms of the nearest-neighbour method, an excellent example of this is the *crowding distance operator*. The crowding distance of a solution is the perimeter of an imaginary rectangle defined by its left and right neighbours and infinity if there is no neighbour. This diversity preservation method is most relevant to this book.

Elitism Elitism is the fourth element to consider in the design of MOEA. It involves the preservation and storing of the best solutions generated during a search for the Pareto optimal set of solutions. In elitism, an *archive* or secondary population is used to store high-quality solutions. When the constituent solutions in the archive do not take part in the search, the elitism is called *passive*. On the other hand, it is known as an *active* elitism if the archive's population is used to generate new ones during the search. Finally, the objective of elitism is to ensure that previously obtained non-dominated solutions would not be lost in successive generations.

After discussing the design of MOEAs, it is crucial to understand how their performance is measured. The next section discusses the performance evaluation of MOEAs.

2.2.10   MOEA performance evaluation

Two key indicators widely used to measure the performance of MOEAs, are the amount of computational resources utilised to find a solution and the quality of the solution. Researchers, have extensively used the amount of computational time it takes an algorithm to reach a good solution, as a yardstick for evaluating the algorithm. This attribute is quite straight forward as it measures the computational run time of the algorithm. In terms of the quality of solutions, various schemes are used by researchers, to either justify the performance of their algorithm or to show that they are better existing ones. Such schemes include:

- Benchmarking Chase et al. (2009); Laumanns et al. (2005) where the obtained result is compared to those of other well-known problems that are similar.

- Comparing the obtained solution with the Pareto optimal set of a small problem obtained using exact solution techniques Arbex and da Cunha (2015). This is only applicable in cases where the Pareto optimal set can be obtained.

- Using performance indicators that measure the convergence, diversity and other features of the non-dominated solutions. Some instances of these indicators are the hypervolume, generational distance, inverted generational distance, maximum Pareto error and spacing or spread (Michalewicz, 1992).

Real-life multi-objective problems like the TNDP are known to be very complex. Additionally, obtaining their Pareto optimal set is near impossible. Hence, there is indeed no way to be sure, that a non-dominated set of solution(s) is globally optimal or how close the set is to the Pareto optimal set. To this end, Talbi (2009b) asserts that despite the existence of these earlier mentioned indicators, *there are no generally accepted standard for measuring the performance of meta-heuristics or the solutions they yield.* In the next section, the SBO literature is reviewed. The review particularly focuses on the previous application of the technique in the transportation.

2.3 Simulation-based optimisation of public transit networks

According to Gosavi (2015b), SBO is a relatively new academic discipline that has only gained traction in operations research literature in the last two decades. It deals with integrating optimisation and simulation to solve complex optimisation problems like the TNDP. While SBO has found application in other areas of transportation research like signal design, there are no published works that show its application to the design of public transport networks. A detailed review of the SBO literature may be found in Alrabghi and Tiwari (2015); Amaran et al. (2016); Arisha and Abo-Hamad (2010); Azadivar (1999); Hachicha et al. (2010). In these works, various schemes are used to classify SBO problems and solution algorithms. Some classifications of the problems are by their input variables (quantitative versus qualitative variables), the number of objectives (single and multi-objective) (Hachicha et al., 2010). Others are by the type of parameter (discrete or continuous), and the optimisation procedure used. In terms of solution approaches, four prominent ones set out in Arisha and Abo-Hamad (2010) are gradient-based methods, statistical methods, meta-models, and metaheuristics, though more are available in Amaran et al. (2016). In terms of transportation systems, only two of these—meta-models and meta-heuristic have found an application in the literature. Meta-model based techniques use an analytical approach to approximate the objective function. A meta-model replaces parts of the simulation model with a mathematical function that mimics the input and output behaviour of the simulation's stochastic response. Doing this reduces the computational burden of the problem as the meta-model can generally be resolved with deterministic optimisation techniques. Public transport-related research that has adopted this solution approach is Osorio and Bierlaire (2013). The author introduced a SBO method that facilitates the efficient use of stochastic urban traffic simulators to address various transportation problems. In the work, a meta-model is used to integrate information from a simulator with an analytical queueing network model. Essentially, the proposed meta-model combines a general-purpose component (a quadratic polynomial), with a physical component (the analytical queueing network model). The resulting SBO framework is computationally efficient and can be applied to complex problems with tight computational budget constraints. The method was also used to evaluate the performance of a traffic signal control problem for the Swiss city of Lausanne, under different demand scenarios. The results show that the average travel times on the network was reduced and more reliable.

Also, in Osorio (2016); Osorio and Selvam (2015) the authors use the SBO technique in evaluating the impacts of change on public transport networks, especially in terms of the network performance. Their approach involved combining multiple stochastic traffic simulators to identify points on a network with high-level performance in terms of a stated indicator. The simulators had different levels of computational efficiency and accuracy. The resulting algorithm was then used to measure the performance of a traffic signal control problem on a small network. The results showed that the algorithm was capable of obtaining signal plans with good performance and with a reduced computational cost.

In SBO solutions involving meta-heuristics, the latter is generally integrated with a simulation. Metaheuristics have been discussed in some details in preceding sections of this chapter. However, they have an advantage over the earlier mentioned SBO methods. This advantage is their ability to find good solutions even when the search space is high-dimensional and not continuous or when qualitative decision variables are involved (Arisha and Abo-Hamad, 2010). A typical application of this approach to public transport systems is Song et al. (2013), who proposed a SBO method for evaluating and optimising sustainable transportation systems. The objective of the work is to get an optimal combination of transport planning and operational strategies that minimise the generalised costs of multimodal travel. Examples of the strategies are automobile congestion pricing and public transport fares. Four major parts of their model are the strategy, simulation, evaluation, and optimisation. The SBO process was also used to validate the efficiency of each strategy. The tools used to implement the model was the traffic simulation software VISUM in combination with a GA. The model has been used to study a small multimodal network in China to show its feasibility.

Similarly, the research in this **book** is a meta-heuristic SBO as it combines a MOEA with ABS. The context of the application, however, differs as the model would be applied to a TNDP. The integrated model involves using an agent-based simulation instead of a trip-based travel demand model within the meta-heuristic optimisation solution framework. The integration entails calling the ABS within the optimisation algorithm repetitively. Multiple runs of the simulation would be executed in each iteration of the meta-heuristic. A mean of the results is then used as input in the reproduction process of the MOEA.

2.4 Summary

The focus of this chapter has been to develop a theoretical framework for this **book**. It describes relevant concepts and locates the **book** within the existing literature. The first section of the chapter introduced the concept of models and simulation, then gives a more in-depth discussion on ABMs. In section two, an extensive review of the TNDP and its existing solution methods is done. The section also discusses the concept of MOO. Subsequently, in section three, a discussion of critical aspects of the SBO literature is first done. This is followed by a brief discussion about the integration of the ABS and MOP to realise the proposed SBTNDM. Looking ahead to the next chapter, the implementation of the ABS is discussed in detail. The next chapter is the first of three consecutive ones, that deal with implementing the SBO transit network design model. Chapter 3 discusses the simulation component. Chapter 4 looks at the optimisation aspect of the solution and lastly, chapter 5 presents the integrated simulation-based optimisation framework for public transit network design.

Chapter 3

Agent-based travel demand simulation

In this chapter, an Agent-based Travel Demand Simulation (ABTDS) called the Multi-Agent Transport Simulation (MATSim) is described. It will serve as the evaluation component of the proposed Simulation-based Transit Network Design Model (SBTNDM). To this end, the discussion centres around the model's operations, data requirements, and how it simulates public transportation networks and services. A rationale for the choice of MATSim is given in the next section.

3.1 Choice of ABM for the proposed research

This section discusses the choice of simulation tool used in the book. It highlights some of the critical differences between MATSim and other agent-based demand models, that make the former most suited for the work. An observed feature in older activity-based travel demand models such as those proposed in Bhat et al.(2004); Pendyala (2004), is that their results are in the form of time-dependent Origin Destination (OD) matrices, which are then fed into a Dynamic Traffic Assignment (DTA) model. The major limitation of this approach is that they are not fully disaggregated, since static OD matrices is still used to represent travel demand. Therefore, they do not capture to a full extent the disaggregated nature of every traveller's behaviour on the transit network. An exception to this is the Transportation Analysis and Simulation System (TRANSIMS) (Bowman et al., 1999; Lawe et al., 2010; Nagel et al., 1999), which generates individual activity plans as input to the DTA. Though a more accurate implementation of an activity-based microsimulation model, TRANSIMS was initially difficult to obtain outside the United States due to its proprietary licensing. This development motivated the production of an open-source and fully disaggregated Agent-based Simulation (ABS) platform known as MATSim. The newer platform offered greater access to transportation stakeholders worldwide. Although, TRANSIMS has become open source in the meantime, MATSim is preferable for this **book**, since it now performs better than the former in the following aspects.

- With the MATSim format, it is possible to do all information exchange between modules with a standardised Document Type Definition (DTD) format, varying only the level of detail of the included information. This means that arbitrary combinations of partial activity-based demand generation modules can be used simultaneously without conflict.

- The traffic flow simulation module in MATSim is more straightforward than that of TRANSIMS. However, the latter runs slower than the former and requires significantly more computing time (Balmer et al., 2006).

- The dynamic traffic assignment process in MATSim keeps track of the uniformity of activity chains along the time axis. This implies that travellers have to spend a precise minimum amount of time at each activity before they can proceed. The same is not the case for TRANSIMS.

- MATSIM is capable of tracking multiple plans for ever agent simultaneously. This is not possible in TRANSIMS as the model would require substantial implementation alterations.

The next section discusses the data requirement for MATSim.

3.2 MATSim data requirement

This section gives a description of the data elements required to set up the MATSim model, and the logical relationships between them. All the data used are based on travel demand and supply data collected from the MyCiTi Bus Rapid Transit (BRT) service in the City of Cape Town (CoCT). The travel supply-side data is obtained in the form of the General Transit Feed Specification (GTFS) data for the service. On the other hand, passenger ridership data or Automated Fare Collection (AFC) represents the demand side data of the BRT service. The Java programming language (Arnold and Gosling, 2000) is used for all data processing discussed in this chapter.

3.2.1 General transit feed specification

Normally, the transit network and operational schedule data are organised in the GTFS data format. This is a standard format that public transportation agencies around the world use to share information with the general public regarding their services. Some of the commonly shared information includes geographical data like routes, stops, and the associated schedules for those routes. GTFS was developed in 2007, through the pioneering collaboration of *TriMet*, a public transportation agency in Portland Oregon, USA and the Google transit team (McHugh, 2013). The primary motivation for this project emanated from the need to make information about public transportation services more accessible to people that need it. Since its launch, the service has become one of the major ways of advertising and promoting the use of public transport services. More people can now easily access information about different public transport services, by mode, from their mobile devices in real-time.

However, to develop this service, the problem of storing and reporting public transportation service-related data in a uniform format had to be surmounted. The process of solving this problem by harmonising the way various public transportation authorities stored and reported their data gave rise to the GTFS. Today, the GTFS data specification makes it possible for stakeholders in the public transportation industry, to uniformly, analyse data from different agencies, without the need for customised data processing as before. Furthermore, in terms of application GTFS, data is now used extensively in trip planning and maps, ride-sharing, timetable creation, data visualisations and real-time transit information in Antrim and Barbeau (2013); Wessel et al. (2017). The GTFS data scheme enables public transportation agencies to publish their data for different uses.

The data format is essentially a set of text files with comma-separated values that represent different fields—agency, routes, stops, trips, stop times, calendar and calendar dates. The individual text files are then, compressed into a *.zip* file called a GTFS feed (Wong, 2013). GTFS data is the major source of transport service-related data such as transit network, schedules and vehicles. Hence, to obtain these data elements, the GTFS feed is parsed or analysed to reveal its components. Subsequently, the required data is written to file. The next section discusses the AFC travel demand data for the MyCiTi BRT service.

3.2.2 Automated fare collection data

The MyCiTi BRT is a closed system with a restricted tap-in and tap-out access to passengers. It uses an AFC data system that deploys smart card technology in the collection and storage of the ticketing and travel data for passengers. The system also determines the fares that travellers should pay for their trips. This usually means that the ticketing process is easier for passengers and revenue collection is more efficient in general. It is also a more affordable way of collecting transit ridership data than conducting surveys. To access the system, each passenger must have a smart card with the appropriate unique identification numbers. The AFC system reads the passenger's smart card to determine that they have sufficient credit for their trip and by so doing a record of valid trips made by the passenger may be stored. The smart card unique number enables people's movement to be tracked anonymously within the system, without violating their privacy or obtaining personal information. This facilitates the automated collection of passenger flow and network utilisation information; such as unique passenger transaction numbers, boarding and alighting locations, bus route interactions with the route name, along with the date and time of transactions. The interactions are typically in the form of bus transactions—*boarding, alighting* and *transfers* at stops.

Historically, this type of information has been used to monitor transit system performance and to support decision-makers as they formulate policy. The trip data is collected for the duration of the daily BRT service operations, then aggregated as monthly reports and stored as spreadsheets. The city of Cape town's Transit Development Authority (TDA) made this data available for use in this **book**. The AFC data is a major input data requirement for the ABS component of the Simulation-based Optimisation (SBO) model. Hence, there is need to pre-process and convert the data into the required input format. The steps taken in pre-processing the data follow in the next section.

AFC data pre-processing

Pre-processing the AFC data involves reading in the data from the storage database and cleaning it. To clean the data, incorrectly recorded data points are deleted. This is done to eliminate inconsistencies in the dataset, which may negatively impact the outcome of any simulations done with the data. The data is then organised in a way that makes it usable in the SBTNDM.

While the AFC can provide a consistent flow of passenger transaction data, it has two significant limitations that affect the suitability of the data for this **book**. The first limitation is the inability of the system to record the unique sequence of each passenger's trip as they occur within the system. This affects the ability of the transport planner to understand the travel path and trip sequence of passengers clearly. The other limitation is the inability to accurately link the trips mentioned above to the MyCiTi network nodes and links. This issue arises, because the stop name encoding on the AFC is different from that used in the stop name information of the GTFS data.

Other inconsistencies within the data occur from unplanned events or disruptions to the system, which adversely impacts the collection and storage of the data. They include loss of Global Positioning System (GPS) signals while recording the data; internal data processing errors (read and write errors), or disruptions in the bus services. These issues may affect the integrity of the collected data. Therefore, pre-processing the data is crucial. The steps taken to pre-process the AFC data are discussed next.

Split the data into daily trips While agent-based travel demand models simulate people's daily travel demand over 24-hour periods, the AFC is stored in monthly batches. Therefore, it is necessary to split the raw data into daily slices that will be used in the simulation. Each monthly data chunk is made up of all the daily transactions for all the days of operation in a given month. The data is ordered by the time of day when the transactions occur. After splitting the data into daily slices, the data is further organised into *trip chains* or *plans*. This is the subject of discussion in the next section.

Chaining of daily trips After splitting the dataset into daily slices, the unique users' id or card transaction number in the slices are identified. Trips are then mapped, to each user id, in the order, the trips are performed by the time of day. The transaction for each user id is filtered according to the type of operation in any of these order boarding-alighting (B-A), connection-alighting (C-A) and boarding-connection-alighting (B-C-A). In the AFC data, connections are trips that continue from the last disembarkation point within a 15 minute duration.

Hence, trips that meet the formats mentioned above are grouped as the trip-chains or plans for the traveller. Any trips which do not conform to this format are discarded. Lastly, the data is checked to ensure that certain constraints are obeyed, like ensuring that the passenger's next trip occurs later than the previous alighting.

Create the plan data This is the final step in the process; it involves matching the valid trip chains obtained in the preceding step with the corresponding passenger. This matched data is then written to file, as the daily *population plan* file that would be used in the agent-based simulation. A flowchart for processing the AFC data can be seen in Figure 3.1.

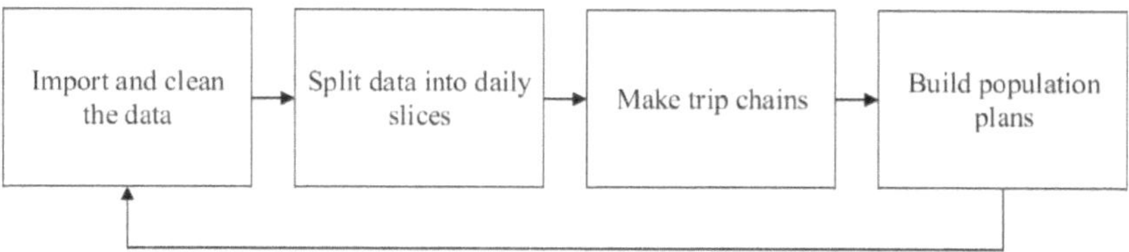

Figure 3.1: Steps for converting the AFC data to input for the Agent-based Model (ABM).

3.2.3 MATSim input data

The input data required for the agent-based travel demand simulation consists of a transit network, transit schedules, vehicle fleet, agent plan or daily activity chain and configuration settings. These are saved as Extensible Markup Language (XML) (Bray et al., 2006) files. The agent's *plan* files are extracted from the MyCiTi AFC dataset, while the transit network, operational schedules are obtained by converting the MyCiTi GTFS feed to the network and operational schedules. The data extraction is done with the aid of ad-hoc or heuristic algorithms developed for that purpose. Lastly, a configuration file is created, which contains parameters and their values that are needed to control the simulation. A brief description of the earlier mentioned components of the input data is given in the next section.

Transit network The network is represented as a list of nodes and links in a file named `network.xml`. The nodes represent stops and stations on the MyCiTi BRT service, while, the links represents the connecting infrastructure that makes up its operational routes. In terms of graph theory, the network can be represented abstractly as a graph $G = (N, E)$. When correctly arranged, the graph is a topology of the network. Each node or stop $n \in N$ on the network is assigned a unique identifier, and other attributes such as the `x` and `y` coordinates of the node's geographic location. Other optional characteristics could be the name of stops if that information is available. Similarly each link $e \in E$ is given a unique identifier alongside other attributes like `origin node`, `destination node`, `length`, `free flow speed`, `mode` of transport that will operate on the link, `number of lanes`, and `link capacity`. Lastly, all links must be uni-directional; if a road is traversed in both directions, two links should be created, one in each direction. See sample network file in Figure 3.2.

```xml
<?xml version="1.0" encoding="UTF-8"?>
<!DOCTYPE network SYSTEM "http://www.matsim.org/files/dtd/network_v2.dtd">
<network>
<!-- ================================================================= -->

    <nodes>
        <node id="MyCiTi_1" x="-53196.450154726146" y="-3755010.0058102254" >
        </node>
        <node id="MyCiTi_10" x="-54879.37761845079" y="-3753903.660850382" >
        </node>
        <node id="MyCiTi_100" x="-46659.23389528884" y="-3749500.821686937" >
    </nodes>

<!-- ================================================================= -->

    <links capperiod="01:00:00" effectivecellsize="7.5" effectivelanewidth="3.75">
        <link id="MyCiTi_0" from="MyCiTi_301" to="MyCiTi_301" length="50.0"
            freespeed="10" capacity="500.0" permlanes="1.0" oneway="1" modes="pt" >
        <link id="MyCiTi_1" from="MyCiTi_301" to="MyCiTi_300" length="498"
            freespeed="10" capacity="500.0" permlanes="1.0" oneway="1" modes="pt" >
        <link id="MyCiTi_10" from="MyCiTi_155" to="MyCiTi_126" length="1003"
            freespeed="8" capacity="500.0" permlanes="1.0" oneway="1" modes="pt" >
        <link id="MyCiTi_100" from="MyCiTi_319" to="MyCiTi_315" length="266"
            freespeed="8" capacity="500.0" permlanes="1.0"
    </links>

<!-- ================================================================= -->
```

Figure 3.2: A typical MATSim transit network file.

Transit schedule This data is presented in a file named `transitSchedule.xml`. The file contains two distinct pieces of information about stop facilities and transit lines on the BRT service. In the first part, the stops are listed, with each stop having attributes like; a geographic coordinate, and a unique identifier (`ID`). It also has a link reference identifier (`refID`) that points to the network link where a stop is located. Each stop can only be served by vehicles operating on the link it references. Other optional properties for stops are a *name* and *blocking* characteristic of the stop. The `blocking` attribute is useful when modelling different types of stops like on-street and off-road bus stops. The second part of the transit schedule file contains information about the transit lines, routes and their operational schedules. These are presented in a hierarchical data structure with each line having two or more routes. The routes, in turn, have a `mode`, `route profile` (ordered list of stops to be served),

network route (links along the route), and a list of **departures** from the stops. See sample transit schedule file in Figure 3.3.

```xml
<?xml version="1.0" encoding="UTF-8" standalone="no"?>
<!DOCTYPE transitSchedule SYSTEM "http://www.matsim.org/files/dtd/transitSchedule_v1.dtd">
<transitSchedule>
  <transitStops>
    <stopFacility id="1" isBlocking="false" linkRefId="180" name="Beng" x="-53196.45" y="-3755.00"/>
    <stopFacility id="1.1" isBlocking="false" linkRefId="212" name="SmithRoad" x="-53196.45" y="-37010.00"/>
    <stopFacility id="1.2" isBlocking="false" linkRefId="307" name="Auburn" x="-53196.45" y="-37510.00"/>
  </transitStops>
  <transitLine id="100">
    <transitRoute id="100_0">
      <transportMode>bus</transportMode>
      <routeProfile>
        <stop arrivalOffset="00:00:00" awaitDeparture="false" departureOffset="00:00:00" refId="166"/>
        <stop arrivalOffset="00:33:00" awaitDeparture="false" departureOffset="00:33:00" refId="29"/>
      </routeProfile>
      <route>
        <link refId="MyCiTi_294"/>
        <link refId="MyCiTi_2005"/>
      </route>
      <departures>
        <departure departureTime="05:00:00" id="10010000" vehicleRefId="0"/>
        <departure departureTime="07:45:00" id="100100011" vehicleRefId="11"/>
      </departures>
    </transitRoute>

        .
        .
        .

      </transitRoute>
  </transitLine>
</transitSchedule>
```

Figure 3.3: A typical MATSim transit schedule file.

Population The population input data usually consists of a group of synthetic agents or travellers and their daily activity plans. The synthetic population used in this **book** corresponds to the users of the MyCiTi BRT system. Their unique smart card number represents each agent or passenger individual in the population. This data is created from the AFC ridership data. The agents have precisely one plan showing their chain of activities and trips during the modelling time frame or the daily operating hours. The plans in turn have **activity** and **leg** attributes. The activities have **type** and **x-y** attributes, which describes the nature of the activity and the geographic location where it is happening. Activities, except for the last one for the day, is assigned an **end time** attribute. The leg attribute describes how the agent travels between activity locations. Each leg is assigned a *mode* of transport with an optional **travel time** attribute. Usually, the agent starts a leg right after the previous activity (or leg) ends. Figure 3.4 depicts a typical population file used in this work.

```
1  <?xml version="1.0" encoding="utf-8"?>
2  <!DOCTYPE population SYSTEM "http://www.matsim.org/files/dtd/population_v6.dtd">
3
4  <population>
5
6  <!-- ====================================================================== -->
7
8      <person id="1">
9          <plan selected="yes">
10             <activity type="home" x="-4858.77" y="-37101.3910760316" end_time="06:00:00" >
11             <leg mode="pt">
12             </leg>
13             <activity type="work" x="-4708.31" y="-37516.19" start_time="08:00:00" end_time="17:30:00" >
14             <leg mode="pt">
15             </leg>
16             <activity type="home" x="-4858.774844986445" y="-37101.39">
17         </plan>
18
19     </person>
20
21  <!-- ====================================================================== -->
22
23      <person id="2">
24          .
25          .
26          .
27
28      </person>
29
30  <population>
```

Figure 3.4: A typical MATSim population file.

Transit vehicle This input data is saved in a file named `transitVehicles.xml`; see Figure 3.5 for a depiction of a sample transit vehicles data file. The first part of the file is called `vehicleType`, which describes the features of the vehicle with attributes like, `capacity`, `length` and `description`.

```
1  <?xml version="1.0" encoding="UTF-8" standalone="no"?>
2  <vehicleDefinitions xmlns="http://www.matsim.org/files/dtd">
3
4      <vehicleType id="defaultTransitVehicleType">
5        <capacity>
6          <seats persons="101"/>
7          <standingRoom persons="0"/>
8        </capacity>
9        <length meter="7.5"/>
10       <width meter="1.0"/>
11       <accessTime secondsPerPerson="1.0"/>
12       <egressTime secondsPerPerson="1.0"/>
13       <doorOperation mode="serial"/>
14       <passengerCarEquivalents pce="1.0"/>
15     </vehicleType>
16     <vehicle id="tr_0" type="defaultTransitVehicleType"/>
17     <vehicle id="tr_1" type="defaultTransitVehicleType"/>
18     <vehicle id="tr_2" type="defaultTransitVehicleType"/>
19
20  </vehicleDefinitions>
```

Figure 3.5: A typical MATSim transit vehicles file.

A transit vehicle may refer to a single-vehicle like a bus, or multiple vehicles coupled together, for example, a train with several wagons. In MATSim, the latter is modelled as one vehicle with a high number of seats. In the second part of the **vehicle** data file, the actual vehicles allocated to the system are listed. Each vehicle is given a unique identifier and must be of a type that has been previously specified in the **vehicleType** attribute. In this work, one transit vehicle is defined for every assigned trip departure in the timetable.

Configuration The configuration data for MATSim is saved in a file known as **config.xml**. The file has parameter settings used to manipulate the simulation. These parameters are defined in modules within the configuration file. Instances of the modules are **plans**, **network**, **controler**, **planCalcScore**. These are responsible for setting the input plans and network files, the number of iterations the simulation should run for, and describing the activity types and duration, respectively. Other modules can be set depending on what the simulation objectives are. See Figure 3.6 for a depiction of a typical configuration data file.

```xml
1  <?xml version="1.0" ?>
2  <!DOCTYPE config SYSTEM "http://www.matsim.org/files/dtd/config_v2.dtd">
3  <config>
4
5      <module name="global">
6          <param name="randomSeed" value="4711" />
7      </module>
8
9      <module name="network">
10         <param name="inputNetworkFile" value="./network.xml" />
11     </module>
12
13     <module name="plans">
14         <param name="inputPlansFile" value="./plans.xml" />
15     </module>
16
17     <module name="controler">
18         <param name="outputDirectory" value="./output" />
19     </module>
20     <module name="planCalcScore">
21         <param name="learningRate" value="1.0" />
22     </module>
23
24 </config>
```

Figure 3.6: A typical MATSim configurations file.

3.3 MATSim

The MATSim is an ABS framework, designed to model large scale transportation
scenarios in very fine details. It can simulate, the microscopic activities of people on
a public transportation network over a 24-hour duration. Conceptually, a MATSim
simulation consists of two layers that are characterised as *physical* and *mental*
in Nicolai (2013) and Rieser (2010). The *physical* layer is also called a mobility or
traffic flow simulation, and it represents the tangible parts of a transit system, like
agents or travellers, their activity plans, activity locations, vehicle fleet, network
infrastructure and other concrete elements of the system.

The physical layer reflects the agents and how they interact on a transit network.
On the other hand, the *mental* layer represents the abstract part of a transit network
system. It describes how the agents receive and process information from the network
environment; and how this affects their decisions and choices on the network, with
the ultimate goal of improving their overall travel experience. For example, if a
traveller who has planned to use a specific transit route, now receives information
about a sudden closure on that route. This will likely make the traveller choose a
different route to their destination. In conclusion, as a transportation planning tool,
MATSim enables public transit authorities and planners, to study the impact of
planned and already implemented transport policies. This is achieved by simulating
various public transportation planning scenarios. Instances of these scenarios include
providing additional bus lanes and re-routing train services to areas, where, they did
not previously operate. Others are increasing the service frequencies of a bus service
or even setting up a road pricing scheme.

The next section describes how public transport is modelled on the MATSim
platform. This is important, since, the evaluation of transit route networks in the
proposed SBTNDM will be done with MATSim.

3.3.1 Public transit modelling in MATSim

This section discusses the operationalisation of public transport systems in MATSim.
The steps taken to build a public transportation scenario on the simulation platform

are also discussed. MATSim organises public transit system data in a format that is commonly used by public transit services worldwide Horni et al. (2016). In the case of the MyCiTi BRT network a line modelled in MATSim will comprise two or more transit routes. The route itself is a sequence of road links that facilitate the MyCiTi buses to run on the route. Each route serves one direction of travel and enables buses to move to and from the depot at the end and beginning of a day, respectively. The routes also have as an attribute the list of **departures**, which gives information about the time a vehicle starts at the first stop on that route. Furthermore, a route includes a sequential list of transit stops that are served, alongside operating timetables, which indicate when vehicles arrive or leave a stop. The times are specified as offsets in time units from the departure at the first stop. At each subsequent stop, the offset is added to the initial departure time at the first stop. Each departure contains a vehicle's start time on the route and a reference to the vehicle. As the timing information is part of the route, it becomes possible to have routes with identical stop sequences but different time offsets. Stop locations are described by their coordinates and an optional **name** or **id**. They must be assigned to unique lines of the network for the simulation. A typical MATSim loop with its steps are shown in Figure 3.7. Subsequently, a description of these steps would be given.

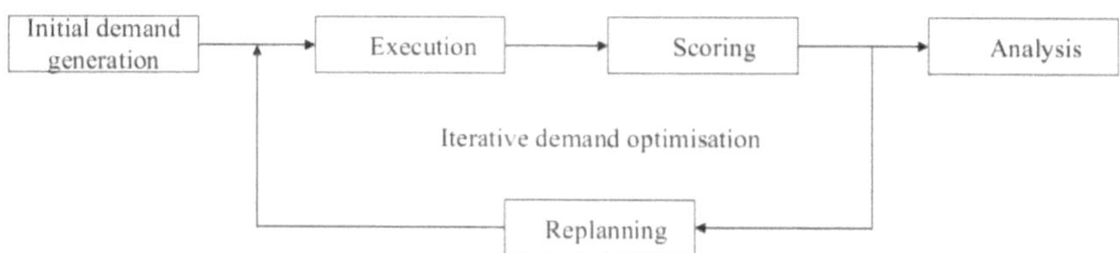

Figure 3.7: MATSim simulation process, courtesy Horni et al. (2016).

Initial demand generation The initial demand is generated by creating daily activity plans, from the socio-economic and demographic data of agents within a given transportation area. The demand is usually generated, through sampling or discrete choice modelling and subsequently converted to activity chains or plans for the agents. The data comprises an inventory of agents with their activity schedule or plan. An agent's plan reflects their intention. Each plan should detail a list of their current and next activity locations, trips or mode choice, and time-based information like activity departure, arrival and duration. Typically, each MATSim agent stores a fixed number of daily plans in their memory.

Execution Execution involves simulating the generated demand. The plans are executed sequentially by time of occurrence. It is also done in a way that respects certain boundary conditions like the closing hour of a shop or the maximum link and flow capacity of a road. The constraint represents the physical infrastructure where the activities and trips will be undertaken (Meister et al., 2010). Another name for this step is mobility simulation, or *mobsim*, for short. When the simulation begins, an agent chooses a single plan. This plan is then executed during the simulation. As the simulation progresses, two factors that the passengers consider are the start time of their activities, and the spaces they occupy in the network as they travel to their respective activity locations. The latter affects how other people's plans are executed. For instance, the occurrence of congestion could delay the travel time of agents that would have used the route. It is this competition for the limited network resources, as travellers go about their activities that inform the co-evolutionary design of MATSim. The overall effect, of the actions of individual agents on a network at a given time, defines the prevailing network condition at that time. It also determines the subsequent choices people will make on the network.

Scoring After executing the agents' plans, the plans are evaluated. The score is obtained, by evaluating a plan using a utility function known as a *scoring function*. This function may be considered an objective function in optimisation terms, or a utility function in econometric terms. MATSim uses the scores to measure and compare the quality of a passenger's plan and determine if it should be dropped or not. It describes a traveller's perception of time spent travelling or engaging in an activity with components like; waiting time, travel time or time spent on the activity.

Generally, time spent on the activity, increases the score, while time spent travelling decreases it. Waiting time, such as when agents are stuck in traffic leads to a loss of points. The mathematical formulation of a MATSim scoring function applies a convention that rewards the agent for performing activities (positive utility) and penalises them for travelling (negative utility). This is consistent, with the generally accepted notion that travelling is a disutility. Therefore, if an agent chooses a plan which allows them to spend more time on their activities—work, education or leisure, their reward or positive utility invariably increases. Conversely, if a longer time is spent travelling, the negative score accumulates.

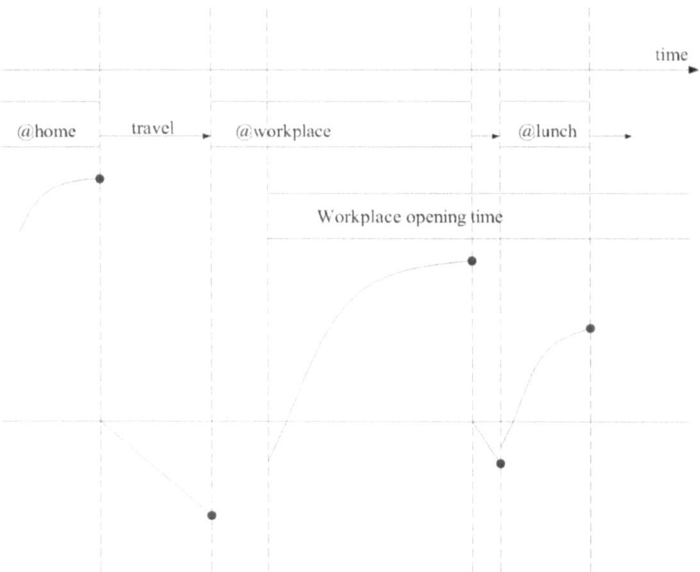

Figure 3.8: Typical process of scoring activities in MATSim, adapted from Horni et al. (2016).

Figure 3.8 describes the scoring procedure in the execution of a typical daily activity plan. It shows that the more activities performed by the agent, the more time spent increases, and so does the *reward or positive score*. The reverse is the case when agents travel rather than perform an activity. The following can be deduced from the graph above: 1) a positive score for the home activity, 2) a negative score for travelling to work. 3) no score for waiting until the workplace opens. Assuming the agent arrives early at the location, 4) a positive score for the work activity, 5) a negative score for travelling to lunch.

The general form of the scoring function, is seen in (3.1), while its constituent variable are presented in (3.2). The function is the total utility derived from performing an activity and travelling to and from the activity. The utility derived from engaging in an activity ($U_{\text{activity},i}$) is a sum of utilities derived from performing the activity ($U_{\text{perf},i}$), arriving late ($U_{\text{late},i}$) or early ($U_{\text{early},i}$) at the activity, and waiting to perform the activity ($U_{\text{wait},i}$). It is reckoned that arriving late is considered a negative utility. However, arriving on time at the activity location does not necessarily increase utility. The second component is the negative utility derived from travelling to an activity location.

$$U_{\text{plan}} = \sum_{i=1}^{N} U_{\text{activity},i} + \sum_{i=1}^{N-1} U_{\text{travel,mode}(i)} \tag{3.1}$$

$$U_{\text{plan}} = \sum_{i=1}^{N} \left(U_{\text{perf},i} + U_{\text{wait},i} + U_{\text{late},i} + U_{\text{early},i} \right) + \sum_{i=1}^{N-1} U_{\text{travel},i} \tag{3.2}$$

The variables in (3.2) are described next.

Performing activities

The utility for performing an activity, which is usually positive, is depicted as:

$$U_{\text{perf},i} = \beta_{\text{perf}} \cdot t_{\text{typ},i} \cdot \ln\left(\frac{t_{\text{perf}}}{t_{0,i}}\right) \tag{3.3}$$

where:

$U_{\text{perf},i}$ = utility of performing an activity.

β_{perf} = the marginal utility for performing an activity.

$t_{\text{typ},i}$ = the typical duration for activities, say eight hours for work.

t_{perf} = the actual time spent on an activity.

$t_{0,i}$ = the duration when utility starts to be positive.

Waiting

The penalty for arriving late at an activity, which is usually negative, is then:

$$U_{\text{wait},i} = \beta_{\text{wait}} \cdot t_{\text{wait},i} \qquad (3.4)$$

where:

$U_{\text{wait},i}$ = utility associated with waiting to start an activity.

β_{wait} = the negative marginal utility of time spent waiting.

$t_{\text{wait},i}$ = the time spent waiting such as for the transit vehicle to arrive.

Arriving late at the activity

The penalty for arriving late at an activity, which is usually negative, is:

$$U_{\text{late.arr},i} = \begin{cases} \beta_{\text{late.arr}} \cdot (t_{\text{start},i} - t_{\text{latest.arr},i}) & \text{if } t_{\text{start},i} > t_{\text{latest.arr},i} \\ 0 & \text{else} \end{cases}$$

$$\qquad (3.5)$$

where:

$U_{\text{late.arr},i}$ = penalty for arriving late.

$\beta_{\text{late.arr}}$ = marginal utility of arriving late.

$t_{\text{start},i}$ = start time for the activity.

$t_{\text{latest.arr},i}$ = latest possible penalty-free activity starting time.

Leaving early from the activity

The reward for arriving late at an activity, which is usually negative, is depicted as:

$$U_{\text{earliest.dep},i} = \begin{cases} \beta_{\text{earliest.dep}} \cdot (t_{\text{end},i} - t_{\text{earliest.dep},i}) & \text{if } t_{\text{end},i} > t_{\text{earliest.dep},i} \\ 0 & \text{else} \end{cases}$$

$$\qquad (3.6)$$

where:

$U_{\text{earliest.dep},i}$ = penalty for leaving early.

$\beta_{\text{earliest.dep}}$ = marginal utility of earliest departure.

$t_{\text{end},i}$ = end time for the activity.

$t_{\text{earliest.dep},i}$ = earliest possible activity end time.

Travelling

The disutility or a penalty for travelling is:

$$U_{\text{trav}} = \beta_{\text{trav,mode}} \cdot t_{trav} + \beta_m \cdot m_{\text{trav}} + (\beta_{\text{dist,mode}} + \beta_m + \gamma_{\text{dist,mode}}) \cdot d_{\text{trav}} + \beta_{\text{transfer}}$$

$$(3.7)$$

where:

U_{trav} is the penalty for travelling

$\beta_{\text{trav,mode}}$ is the marginal utility of travelling by a given mode

t_{trav} is the time spent travelling

β_m is the marginal utility of money

m_{trav} is the change in monetary value of a fare caused by the travel

$\beta_{\text{dist,mode}}$ the marginal utility of the distance travelled by a given mode

$\gamma_{\text{dist,mode}}$ cost per unit distance travelled

d_{trav} is the distance of the leg

β_{transfer} is a transfer penalty e.g. incurred in using transfer points on transit systems.

Calibrating the scoring function

The main goal of calibrating a travel demand model is to ensure that the values of the model's parameters fall within acceptable and realistic values. Validation, on the other hand, ensures that the outcome of the simulation is as close as possible to what is obtainable in the real world. The calibration of the MATSim model in this research, will seek to ensure that the passenger counts obtained in the AFC are comparable with the simulated values. Primarily, the model's parameter values will be adjusted until the predicted travel demand matches the observed data for the base year 2015. The calibration function in (3.8) may be used to achieve this.

$$P(i|y) \exp \cdot \left[V(i) + \sum_{a_k \in i} y_a(k) - q_a(k) \ / \ \sigma_a^2(k) \right] \tag{3.8}$$

where:

$P(i|y)$ is the previous plan choice distribution given y

$V(i)$ is the standard score of a plan i

$y_a(k)$ is the actual observed count at a given location during a specified time k

$q_a(k)$ is the simulated count at given location during a simulated time k

$\sigma_a^2(k)$ is the inverse weight of the measurement

The following steps will be taken to calibrate the model:

- The calibration process starts by registering observed counts at stops.

- After a given simulation iteration, the calibrator works out the utility correction per agent that influences the plan choice to match the count's reproduction in a more efficient way. This choice depends on the calculated utility correction.

- The calibrator is then informed about the network loading conditions after the simulation iteration.

- Calibration effect on the general model can then be adjusted. Instead of a specific hardcoded setting, the configurable weight can regulate the strength of the calibration with the other scoring parts.

Nagel (2011) advocates that the procedure should be kept as simple as possible. This is to ensure that realistic travel demand model prototypes are obtained with a small amount of time and computational resources. Röder et al. (2013) advise that the volume of pedestrian should also be scrutinised closely when the model is calibrated to obtain realistic pedestrian flows. Further suggestions on how to calibrate the model can be found in Rieser et al. (2015).

Replanning or innovation strategy When agents adapt their plans, in response to changes in the transit network, it is known as *replanning*. This is the main *innovative* component of MATSim, as it allows the agent to modify their plans as they *learn* about prevailing network conditions. This innovation enables the agent to maximise their experience on the public transport network. Also, this has a link with travel behavioural changes on the network. Such as when a traveller opts to use a different public transit route if it becomes apparent that their initially selected route will be disrupted. Strategies have to be defined that reflect the agents' course of action in terms of modifying their plans.

The strategy adopted by the transportation planner, or authority, also reflects the overall goal of the simulation. Two types of strategy modules available in MATSim are the *selector* and *innovative* strategy module. The former defines the criteria that would be used by agents to select a plan, while the latter specifies how the selected plan will be modified. Some selector modules include **BestScore, ChangeExpBeta, KeepLastSelected, SelectExpBeta, SelectRandom**. The innovative modules are route innovation—**ReRoute**, time innovation—**TimeAllocationMutator** and mode innovation—**ChangeLegMode**. Details of these can be found in Horni et al. (2016).

In a MATSim scenario, the strategy module is usually a combination of a plan selector and zero or more innovative strategy modules. Two cases described in Rieser et al. (2015) are when no strategy module elements are set, and when at least one strategy element is selected. In the case of the former, there is only a *plan selector* strategy module. Therefore, the plan selected by the agent does not undergo any modification. Conversely, if at least one strategy module element is set, the agent's chosen plan will be copied. The copy is then added to the persons' set of plans, and the new plan is set as *selected* plan. This newly selected plan is, in turn, assigned to the strategy module elements for modification. Each strategy element is assigned a number and weight. The latter is used as the probability that the action represented by the element will be utilised.

Termination and post-analysis MATSim specifies a termination criterion to enable the simulation stop once the condition is met. Meister et al. (2010) describe this termination point as an agent-based stochastic user equilibrium. The system runs until the score of the agent's plan does not meaningfully improve, marking the end of the simulation. The output at the end of the simulation is an *events file*. Every action taken by the agents during a MATSim simulation is recorded as an *event*. Events are pieces of information describing the action of an object on the network at a specific time during the simulation Zilske (2016). The events include starting a trip, boarding or alighting a vehicle or changing to a different route. By default, the time when the event occurred is also recorded. Post-analysis involves collecting and aggregating network performance indicators (passenger mileage, average trip duration, and distance per mode) from the events file. This is done to gain insight into the travel demand and simulated behaviour of agents within the study area.

3.3.2 MATSim scenario for the SBTNDM

A MATSim simulation comprises a scenario that is developed from the input data of a specific public transport system. The scenario used in this **book** is that of the City of Cape Town's MyCiTi BRT network. Building and running the scenario involves preparing the input data for a description of the data. The configuration file, which is one of the inputs contains all the modules and parameter and settings required to run the simulation. The parameter settings for modules like; the `planCalcScore`, `innovation strategy`, and `controler` are critical for the simulation to run successfully. This is because they reflect the overall strategy that is adopted in a bid to improve peoples' travel experience on a public transit network. The scenario's parameter settings are described in this section. Moreover, the scenario is then used in the proposed SBTNDM.

The main parameters in the `planCalcScore` module include the scoring function and a description of the agent's activities. The latter has defined attributes, like the start time, minimum duration activity and closing time of the activity. In these preceding sections, the main activities are defined as vehicular boarding, alighting and connections. The MATSim scoring function was developed by Horni et al. (2012) based on the work of Charypar and Nagel (2005). The creators of MATSim have calibrated this function. Hence, some default values are recommended for use in the scoring function. However, these values may be calibrated further for specific case studies. In this work, the scoring function parameter values combine some default MATSim values with those obtained from the detailed calibration of the Cape Town BRT scenario presented in the work by Willenberg (2018). The MATSim default penalty of -6 stated as the marginal utility of walking, only encouraged travellers to walk rather than use the transit service. Based on the outcome of calibrating this parameter, the penalty was doubled in the scenario to restore the expected behaviour observed in the ridership data. The `controler` module is responsible for setting the number of iterations each simulation will run. Therefore, to guarantee that the best possible solution will be obtained, some experiments have been conducted by varying the number of iterations between 1 and 100. Figure 3.9 shows the optimal number of iterations for this model.

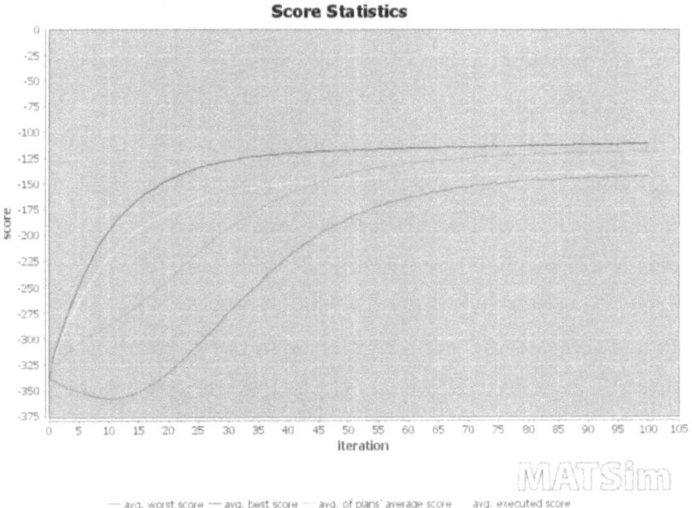

Figure 3.9: Optimal number of MATSim iterations for the SBTNDM;convergence occurs after 80 iterations.

The image show that convergence occurs after 80 iterations. Lastly, the `strategy` module defines how agents select and modify their plans. In this work, each agent is allowed to store a maximum of five plans in memory. This is set with the `maxAgentPlanMemorySize` parameter. A `BestScore` parameter is also defined, which ensures that the plan selected for execution, is the best of the five stored plans. The selected plan may be modified, with the innovation modules at three possible dimensions—mode, time and route. In this work, a route innovation is implemented with the `Reroute` parameter, which allows a traveller to change their route based on network congestion.

3.4   Summary

This chapter discussed the ABS as one of the main components of the proposed SBTNDM. A case was made for MATSim as the simulation platform of choice in the work. Hence, the chapter focused on giving a full description of the MATSim simulation and how it models public transport services. This is followed by the description of a MATSim scenario that is developed for the MyCiTi BRT service. It will be used to evaluate the network solution alternatives in the SBTNDM. This chapter makes it possible to fully understand ABS and how it will function within the SBTNDM. The next chapter discusses the network optimisation component of the SBTNDM.

Chapter 4

Network optimisation

This chapter discusses the multi-objective transit network optimisation algorithm. A Multi-objective Evolutionary Algorithm (MOEA), known as Non-dominated Sorting Genetic Algorithm (NSGA-II), is adapted for this purpose. The NSGA-II is a typical multi-objective Genetic Algorithm (GA). Generally, GAs are meta-heuristic search procedures that can be used to find efficient solutions to optimisation problems like the design of public transport networks. Over the years, GA-based models, have become one of the most efficient methods for solving the TNDP (Buba and Lee, 2018; Kepaptsoglou and Karlaftis, 2009; Nnene et al., 2019b, 2017). In the literature, they are classified as *bio-inspired* algorithms because their operations mimic the principle of natural genetics. They work by enabling the realisation of newer and presumably better generations of solutions from existing ones. A typical genetic algorithm framework consists of a population of solutions or chromosomes. Each chromosome is made up of genes that depend on the particular representation of the chromosome. Furthermore, the algorithm has operators which differ depending on whether the GA is a single or multi-objective one. The actions of these operators on the current population of individuals give rise to offspring solutions, which are generally assumed to be fitter or perform better than their *progenitors*. Multi-objective Genetic Algorithms (MOGAs) require more operators than single-objective GA. This is because the multiple non-dominated solutions and objectives introduce further complexities, which cannot be sufficiently addressed by the traditional process of reproduction. In its operations, the NSGA-II combines both traditional single objective genetic operators with other multi-objective ones.

The rest of this chapter starts by describing why the NSGA-II is chosen as the optimisation algorithm in the book. After this, a description of the algorithm, its operators, and how it can be used to design public transport networks follows.

4.1 Choice of NSGA-II

A major reason for using a MOEA algorithm in this book is that these type of algorithms are better suited to solve problems like the Transit Network Design Problem (TNDP), than traditional optimisation methods like linear programming and gradient search. This is because MOEAs can find approximate solutions that are usually considered acceptable, given the relatively smaller amount of time they use to find the solution. More specifically, the NSGA-II is selected because of its widespread application in the literature. Comparative studies done in Eckart et al. (2000) and Herbawi and Weber (2011), show that the NSGA-II outperforms many other MOEAs. In Eckart et al., for example, eight algorithms are compared, using test functions that address different attribute features of Multi-objective Optimisation Problem (MOP). The attributes are multi-modality, convergence, diversity and others. Herbawi and Weber compared the performance of different evolutionary algorithms in a transportation context, involving a multi-objective route planning problem for ride-sharing. Both studies show that the NSGA-II either match or outperforms most of the other algorithms.

4.2 Data requirement

The only input data required for the multi-objective optimisation algorithm in this work, is a set of feasible transit network routes. The set usually defines the search space for the problem. The NSGA-II selects the initial population of networks to be optimised from this set. In this book, transit networks are created from the existing routes of the MyCiTi Bus Rapid Transit (BRT) network.

4.3   Non dominated sorting genetic algorithm-II

Deb et al. (2000) is credited for the development of the NSGA-II. According
to Knowles et al. (2008); Srinivas and Deb (1994), its precursor, the Non-dominated
Sorting Genetic Algorithm (NSGA), had three limitations which prompted the
development of the algorithm. The earlier mentioned limitations include; high
computational inefficiency, lack of elitism and the need to explicitly specify a *sharing
parameter* (σ_{share}) that guarantees diversity among the solutions. Due to the fact,
that the number of optimal solutions obtained for a problem, depends on the so-called
sharing parameter, an exact guess of the correct sharing parameter for a problem
is not always feasible, hence, according to Deb (2001), the use of the parameter
is problematic. These issues have been addressed by reducing the computational
complexity of the NSGA-II from polynomial power to quadratic, thereby increasing
its efficiency. Elitism, which is known to speed up the performance of the GA
significantly and help prevent a loss of good solutions that have already been
found (Zitzler et al., 2000, 2004, 2002), has also been built into the algorithm. Lastly,
a crowded comparison operator is used to guide the selection process. It combines
the dominance rank (fitness assignment method) and the local crowding distance
(diversity preserving nearest neighbour method) to guarantee a uniform spread of
solutions on the Pareto front. The next section elaborates further on the NSGA-II
operators. A good understanding of these operators will give clarity to the role they
play in the proposed Simulation-based Transit Network Design Model (SBTNDM).

4.4   NSGA-II operators

4.4.1   Non-dominated sorting

The non-dominated sorting is used to *group* or *rank* the solutions into different
Pareto fronts, in the search space of a MOP. Recall that, in the context of an
optimisation problem, a solution is considered to be non-dominated if it performs
better than other solutions on one objective and is not worse in the other objectives.
Usually, the solutions are grouped according to their dominance. Then, the identified

non-dominated solutions are ranked. First, all non-dominated solutions in the population are given a rank of 1. These are then temporarily removed from the population. The next, non-dominated solutions are identified and assigned a rank of 2. This process continues iteratively until the population is empty. Figure 4.1 shows the outcome of this procedure — multiple non-dominated sets by rank.

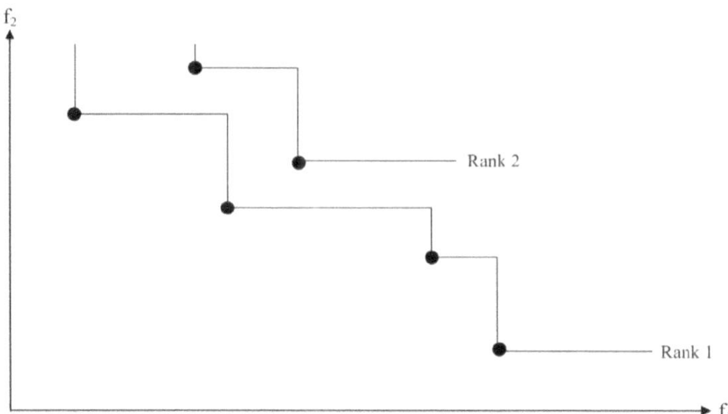

Figure 4.1: Depiction of the non-dominated ranking.

4.4.2   Crowding distance assignment

After the non-dominated sorting discussed above, there may need to identify the best solutions within a given rank of non-dominated solutions. In such cases, the crowding distance operator is used to achieve this. Mainly, the operator is used to rank solutions within a rank. It evaluates the density of solutions within the search space by measuring the distance between a solution s and its n nearest neighbours. For any individual solution, the crowding distance is the perimeter of a rectangle defined by its left and right neighbours. The perimeter is valued as infinity if there is no neighbour. Larger crowding distance values are better as they indicate a greater distance between solutions, hence, a better spread over the entire Pareto front. In Figure 4.2 the non-dominated solutions labelled as rank 1 in the preceding section is isolated to demonstrate the crowding distance operator. The solutions a and d have the best score in terms of crowding because they have higher perimeter values.

This is followed by solutions b and c because the rectangle associated with both solutions is smaller. There are also instances where both the non-dominated sorting and crowding distance operators are combined to work as one operator known as a *crowded comparison operator* see Deb et al. (2000).

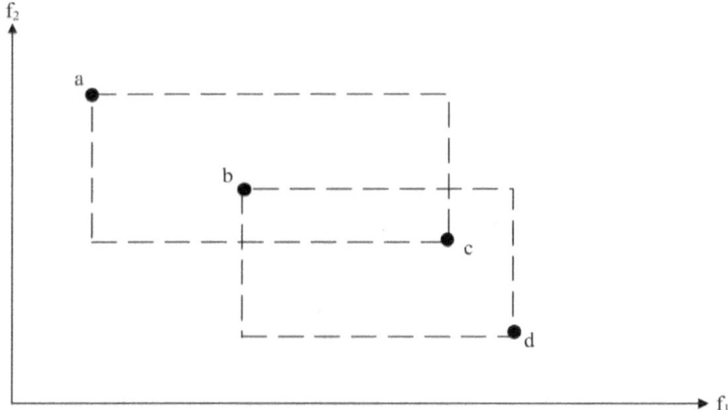

Figure 4.2: Depiction of the crowding distance assignment.

4.4.3 Selection

This is the first of three operators used in the reproduction strategy of GAs. The other two are called *mutation* and *crossover*. The selection operator is used to choose individuals or *parents* from the population that will be used to create the next population or *offspring* in the subsequent generations. The selection is made based on the fitness of each network in the population. The selected individuals are referred to in the literature as a *mating pool*, see Kumar and Jyotishree (2012). The mating pool contributes their characteristic, towards the evolution of the next generation of solutions. When the selection, is not appropriately done, as to reflect the diversity of the parent population, a bias for either the fittest or weakest individuals may arise. The effect of this could be a quick convergence to a local solution in the case of the former or a prolonged convergence rate in the latter scenario. Hence, the effectiveness of the search for an optimised network solution can be impacted either positively or negatively by the selection operator. In this work, a *binary tournament*

80

selection operator (Deb and Bhushan Agrawal, 1995) is used, as is commonly used in the NSGA-II algorithm. The operator chooses two parents randomly and selects the one with a higher fitness score. This process continues until the required number of parents are selected. This selection strategy ensures that only the fittest are chosen for reproduction.

4.4.4 Crossover

The crossover operator generates an offspring solution by mating two designated parents, which are selected randomly from the population. Each parent *donates* a specific number of their constituent routes with those from the other parent, ensuring that the offspring contains routes from both parents. The rate at which crossover occurs is determined by a parameter known as the *crossover probability*. The number of individuals in a population that will undergo crossover is obtained by multiplying the crossover rate by the population size. This implies that if the crossover probability is 0.7, and the population size is 100, 70 individuals within the population will go through crossover. Siriwardene and Perera (2006) observe that a high crossover probability encourages a good combination of parent solutions. This would guarantee that offspring, actually inherit the attribute of the parent network. Despite the optimum crossover probability value being problem-dependent, different authors have proposed some typical values for the crossover probability. De Jong (1975) proposed 0.6, while Grefenstette (1986) proposed a crossover probability of 0.95. Lastly, Schaffer et al. (1989) suggested that typical values lie between 0.75 and 0.95. Types of crossover available in the literature include single-point, multi-point and uniform crossover methods. A single-point (1-point crossover) crossover operator is used in this research. The 1-point crossover randomly selects a corresponding point on both parent networks and swaps every genetic material(routes) between the points, thereby creating an offspring network solution. The crossover process is illustrated in Figure 4.3.

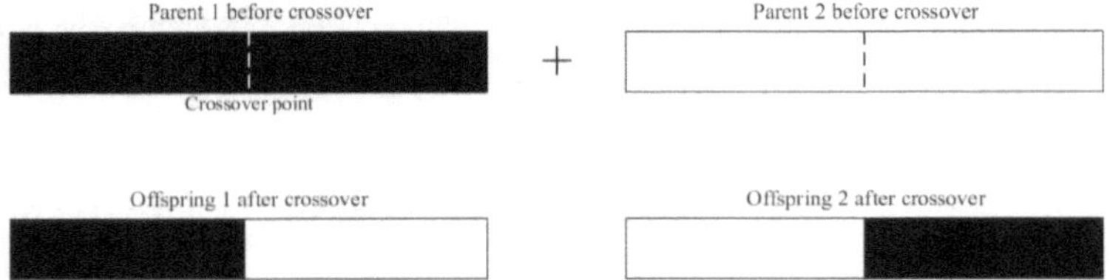

Figure 4.3: An illustration of the crossover process.

4.4.5 Mutation

This is the last operator in the genetic algorithm reproduction process. It introduces diversity into the population by altering the configuration of some networks within the parent population. This is done by randomly identifying the networks to be altered in the first step, then replacing some routes within them with those generated from outside the solution search space (from the study network). This helps to prevent premature convergence by reducing an excessive cluster of too similar routes. The *mutation probability*, just like the crossover probability, determines the rate at which mutation will occur. This means that if a sample solution space has a population size of 20 and each individual in the population has a network size of 10 routes, a mutation rate of 0.1 implies that on average 20 routes in the total populations will be mutated. This is obtained by $20 \times 10 \times 0.1$ (Siriwardene and Perera, 2006). Holland (1992) reports that a high mutation probability increases the possibility of searching more areas in the solution search space. However, it reduces the possibility of convergence to the global optimum solution. The mutation process is illustrated in Figure 4.4.

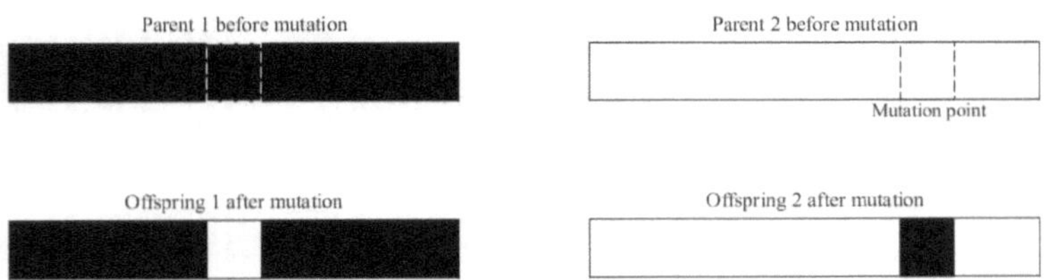

Figure 4.4: An illustration of the mutation process.

4.4.6 Variable encoding

In a typical MOEA algorithm like the NSGA-II, constituent solutions are known as *chromosomes*. Each chromosome, in turn, contain genes. Variable encoding deals with how to represent individual genes within the algorithm. In the TNDP, transit networks are the solutions or chromosome, and their constituent routes are the genes. Usually, some operators of an MOEA perform their actions on the gene, while, others do so on the chromosome itself. Hence, the algorithm will switch between representations of the gene (*genotype*), and the chromosome (*phenotype*), based on, which operator is in use (Sumati et al., 2013). Therefore, it is essential to define a unique representation for the gene level operations, which is different from that of the chromosome. The encoding depends on the nature of the problem, and it must be suitable and relevant to the problem. Moreover, the efficiency of any given encoding is a function of the search operators in use. The primary considerations while defining an encoding scheme are how to evaluate the solution and how the search operators, will manipulate the solution. Typical encodings used in the literature include; bits, numbers, trees, arrays, lists, binary, hexadecimal encodings, among others (Sivanandam and Deepa, 2008). In this **book**, an innovative and customised encoding is used, which is based on a Java Script Object Notation (JSON) data structure (Crockford, 2011). This representation is markedly different from other works in the literature as it accommodates the encoding of each network with its detailed operational schedule.

4.5 NSGA-II operation

A flow chart for the NSGA-II, operations can be seen in Figure 4.5. The process starts with initialising a population of solutions, or chromosomes that serves as the first parents. Subsequently, the solutions are scored or evaluated against the objective functions. Next, a non-dominated sorting procedure is used to rank the population into different Pareto fronts or solutions set. Fitness values equivalent to their ranking are then assigned to each front in ascending order, with the best front ranked as 1 and the next front ranked as 2.

After this, a *binary tournament* selection operator and a *crowded comparison* operator, are used to select parents that will be used to reproduce the offspring. The binary tournament selection randomly chooses two solutions, determines the fitter of both solutions, then adds that one to the mating pool. The crowded comparison operator combines the non-dominated sorting with a crowding distance operator. It is used by the binary tournament selection operator to make the actual comparison between the randomly chosen solutions. First, the solutions are compared based on their dominance. Typically, if the solutions belong to the same front, i.e. they do not dominate each other, the crowding distance is then used to obtain the better solution.

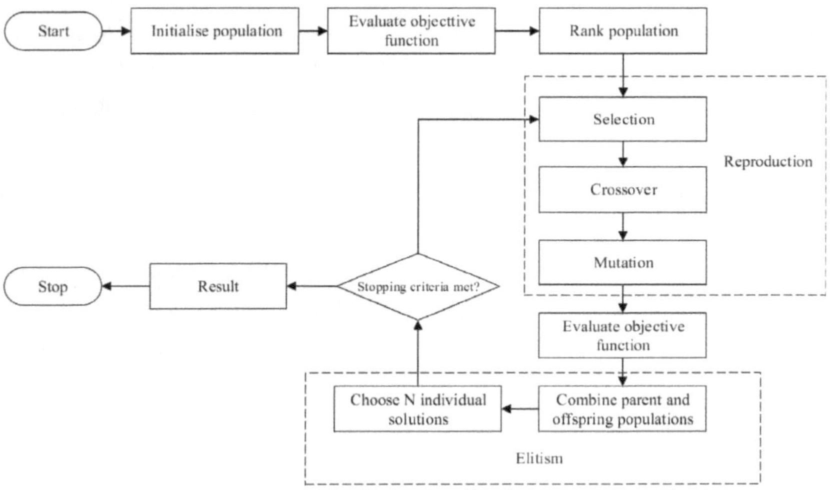

Figure 4.5: NSGA-II flowchart, adapted from Deb (2001).

The next step involves using the crossover and mutation operators to create a population of children/offspring of a size equivalent to that of the first parent. After that, the procedure is slightly different from the first generation. The generated offspring and parent are merged to form a population that is twice the size of the original population in every subsequent generation. The merged population is evaluated and again ranked according to the non-dominance and crowding distance criteria. The better performing half of the merged populations are selected as the new parent population. This process then goes on iteratively until a specified termination condition is satisfied. Elitism is introduced in the algorithm by archiving a small percentage of the best performing or *elite* solutions from both the parents and offspring populations during successive generations. These are reused as part of the parent population in the next generation.

4.6   Using the NSGA-II for transit network design

In designing public transport networks with the NSGA-II, the initial population is a set of feasible network alternatives or chromosomes. The networks generally have different configurations and other attributes. Therefore, the task is to find a network and its attributes among the alternatives, which best addresses the stated optimisation goals. Furthermore, each chromosome possesses a gene. Routes in each network solution represent genes in this context. The best performing chromosome or network in the population represents a globally optimum solution. However, it should be pointed out that for very difficult problems like the TNDP, it is not feasible to know if a solution is the *global optimum*. This is especially true in a multi-objective context where one seeks a set of non-dominated solutions (Pareto optimal front) rather than a single solution. Therefore, an efficient, locally optimal front that is obtained within a reasonable time frame is generally considered acceptable.

To solve TNDPs with the NSGA-II, the chromosomes or networks needs to be encoded in a way that is amenable to the algorithm's operators. Historically, in the literature, string and binary representation are the most common representations used when solving the TNDP. In Buba and Lee (2018), a string is used to represent the network route, while a tuple is used to represent the route's operational frequency as the number of vehicles operated per hour and the unique identifier for that route. However, in this **book**, the earlier mentioned JSON encoding is used. This representation therefore enables genetic operations to be carried out directly on the candidate networks and their detailed operational schedules. This, in turn, allows for the simultaneous handling of the route network design and frequency setting problems. Details of this encoding and how it is used in the proposed SBTNDM is discussed fully in the next chapter, as part of the implementation of the proposed network design solution.

4.7 Summary

In this chapter, the multi-objective route network optimisation component of the proposed Simulation-based Transit Network Optimisation Model (SBTNOM) is discussed. First, a suitable algorithm is selected, followed by a detailed explanation of its operators and how they perform their operations. The chapter ends with a description of how the NSGA-II is used for the actual design of transit networks.

Chapter 5

The simulation-based transit network optimisation model

This chapter focuses on the implementation of the Simulation-based Transit Network Optimisation Model (SBTNOM). It discusses the main steps taken in setting up the model. The next section discusses the data interactions that underpin the SBTNOM. The term SBTNOM is used interchangeably with Simulation-based Transit Network Design Model (SBTNDM) throughout the chapter.

5.1 Data model and interactions

This section briefly discusses the data model for the **book**. It describes the logical interactions between the input data required in the critical elements of the proposed SBTNOM, namely the Multi-Agent Transport Simulation (MATSim) and Non-dominated Sorting Genetic Algorithm (NSGA-II). Firstly, a set of public transit route networks are initialised in the NSGA-II framework. The algorithm has the job of finding the most acceptable solution alternative from the initialised set. However, to do this, the alternatives have to be evaluated. The evaluation is jointly carried out by the optimisation and simulation sub-models. Hence, the primary data interaction, between these components of the SBTNOM occurs during the evaluation of the network solutions. The exchange between these data elements defines the framework that supports the functionality of the SBTNDM. To evaluate the performance of each route network, the `transitSchedule.xml` file containing the routes must be set as an input in the simulation. Each time a route network in the optimisation sub-model is to be evaluated, it is sent to the simulation sub-model. Users' travel demand is then simulated on the network, and the result is analysed and sent back to the optimisation sub-model. The SBTNOM's data interaction is depicted in Figure 5.1.

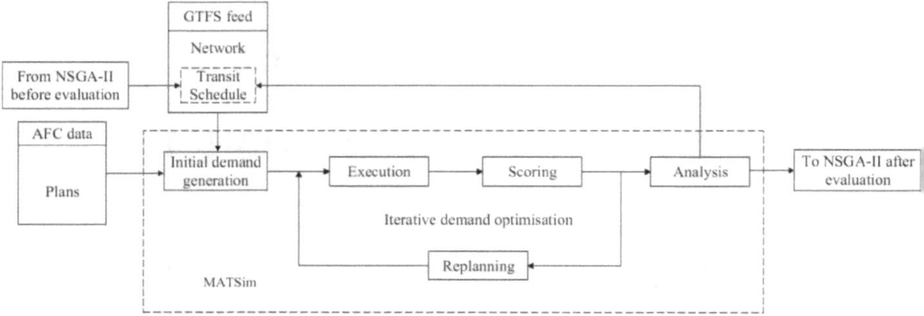

Figure 5.1: The SBTNDM's data model showing different data elements and how they interact.

5.2 Modelling assumptions and limitations

Some assumptions are made in setting up the agent-based travel demand simulation aspect of the Simulation-based Optimisation (SBO).

1) A complete trip or satisfied demand may be in two forms; *boarding–alighting* (B–A) or *boarding–connection–alighting* (B–C–A). The former is a direct trip without transfer, while the latter is a trip satisfied with one transfer required. This specification aligns with how demand coverage is defined in this **book**; demand that is satisfied with zero or one transfer. Only trips satisfied with a maximum of 1 transfer are considered, because as the number of transfers on a trip increases, commuters generally find that trip less attractive. This would lead them to search for alternative routes that are more direct or even change their mode of travel in some cases Owais (2015).

2) In agent-based travel demand models, the demand is generated from people's activities at different locations based on various land uses, such as work, shopping, recreation and education. However, in this work, the network stops and stations, are considered as activity locations. This is due to the availability of Automated Fare Collection (AFC) data and the inability to obtain information concerning other activities or activity locations. Consequently, activities strictly refer to transit network transactions such as passenger boarding, alighting and transfers.

3) One limitation of the SBTNOM is that travel demand is fixed. This entails that demand is based on the existing public transit users alone. Therefore, it will not possible to attract travellers from other modes of transport. This limits the resulting network solution from the optimisation exercise to the MyCiTi network alone rather than one that is more reflective of the entire transit system in Cape Town.

4) The application of the proposed model to a unimodal context — the MyCiTi Bus Rapid Transit (BRT), also, reduces the route choice of the travellers to route choices on only the MyCiTi service rather than between other modes as is the case in a fully integrated public transport system. Therefore, it will not possible to attract travellers from other modes of transport. However in Cape Town where this work is applied, Salazar Ferro et al. (2013) highlights that at present there are no multi-modal public transport systems in the city. Though efforts are being made towards integrating the various services in the city, the possibility of having fully integrated multi-modal networks in Cape Town is still a futuristic prospect. To this end, the model in this work may be considered as a first step taken in the direction of developing a simulation-based network design tool for integrated transit networks in Cape Town.

The next section discusses the solution framework for the implementation of the SBTNOM.

5.3 Solution framework

The solution framework includes a mathematical model of the problem, and the concrete steps taken to implement it. In terms of model implementation, three steps are required to realise SBTNOM. These are 1) a route network generation stage, 2) a route network evaluation phase, and 3) a search for the optimal solution. In the first stage, a heuristic Network Generation Algorithm (NGA) is used for the generation of initial candidate transit networks, which populate the solution search space. They will also serve as parents, from which better offspring networks are produced. The Network Evaluation Procedure (NEP) is executed in the second stage. It is used to evaluate or score the quality of each generated transit network. The MATSim plays this role. In the last phase of the model, a Network Search Algorithm (NSA) based on the NSGA-II is executed to find the best performing network option.

5.3.1 Mathematical model

The network is represented as a graph $\mathbf{G} = (\mathbf{N}, \mathbf{E})$, which is a multiple connection of a finite sets of $n \in \mathbf{N}$ nodes and $l \in \mathbf{E}$ links. The objective functions in (5.1), represent the costs that accrue to two major transit network stakeholders; users as in (5.2) and operators as seen in (5.3). Transit users generally view *generalised cost* in terms of their total travel time (access, waiting, and in-vehicle travel times plus transfers where applicable). On the other hand, operators are concerned with the total operational cost comprising of the total distance and time operated. Operational distance is the cost that accrues from the wear and tears on the operators' vehicles as they traverse the designated routes to satisfy passenger demand. It is typically measured in kilometres. However, the operational time consists of personnel cost element like salaries that accrue throughout operations. Therefore, by minimising these objective functions, the total cost incurred on the network will be optimised for the earlier mentioned stakeholders.

$$Min : Z_1, Z_2 \tag{5.1}$$

$$Z_1 = \beta_{\text{time}} * \sum_{r=1} t_r q_r \tag{5.2}$$

$$Z_2 = \beta_{\text{dist}} * \sum_{r=1} l_r f_r + \beta_{\text{op}} * \sum_{r=1} t_r f_r \tag{5.3}$$

subject to agent-based stochastic user equilibrium on the network:

$$q_r^n = \tau(c(x\{q_r^n\})) \tag{5.4}$$

and some feasibility conditions on route length, frequency and vehicle fleet:

$$l_{min} \leq l_r \leq l_{max} \tag{5.5}$$

$$f_{min} \leq f_r \leq f_{max} \tag{5.6}$$

$$r_{tot} \leq \mathbf{R} \tag{5.7}$$

Where:

$\mathbf{R}$ = set of network routes (-);

r = route on the network (-);

Z = objective function (-);

z_1 = user cost objective function (-);

β_{time} = monetary unit value for user travel time ('000);

t_r = travel time on route r (hr);

q_r = travel demand on route r (pax);

z_2 = operator cost objective function ('000);

β_{dist} = monetary unit value for vehicle mileage ('000);

l_r = length of route r (km);

f_r = frequency on route r (veh/hr);

β_{op} = monetary unit value for vehicle operating time ('000);

q_r^n = individual agent demand on the route r (pax);

n = index of the agent (-);

τ = agent-based probabilistic route choice model (-);

$c(x)$ = network costs (-);

$\{q_r^n\}$ = set of all individual agent route demands on the network (-);

l_{min} = minimum route length (-);

l_{max} = maximum route length (km);

f_{min} = minimum frequency value (veh/hr);

f_{max} = maximum frequency value (veh/hr);

r_{tot} = number of designed routes (-);

The objectives are subject to an agent-based Stochastic User Equilibrium (SUE) model (Horni et al., 2016), which describes the individual traveller's behaviour on a public transportation network, and represented by (5.4). This way of modelling people's travel behaviour extends the conventional stochastic user equilibrium. This is because, rather than aggregating route based passenger volumes as aggregated productions and attractions, as it is done in conventional Trip-Based Models (TBMs), every individual traveller's demand and behaviour is modelled in this case. Furthermore, the route and mode choices used in the traditional user equilibrium is broadened to include other dimensions such as destination choice. Lastly, passenger demand is loaded onto the network, with stochastic algorithms that use time-dependent trip departure times.

In the description of this model it is important to highlight that the route network and its service frequency are used as the problem's decision variables. In the Transit Network Design Problem (TNDP) literature, a decision variable is a resource which is subject to the transit stakeholders' choice in terms of its allocation Curtin (2004). The limits or bounds of their availability is usually defined by a feasibility constraint. Feasibility constraints are parameters that define the limiting conditions of the decision variable(s) in a TNDP. They generally define the feasibility of the optimization problem and ensure that solutions are obtained within reasonable resource limitations.

The feasibility constraints for the model are those on route length, frequency and the vehicle fleet size seen in (5.5) to (5.7). These constraints are used to set the allowed limiting conditions for the allocation of resources on the transit network. Equation (5.5), which is a route length constraint, is introduced to define the upper and lower bounds outside which it would be illogical to operate a bus service. Usually, public transit operators will not run a service on routes that users may conveniently traverse by walking. They also avoid developing excessively long routes (Cipriani et al., 2012). Such routes make schedule adherence difficult or may result in the need to provide too many transfers, which users find unattractive in transit services (Walker, 2011).

Equation (5.6) is a feasibility constraint on transit service frequency. The constraint represents the maximum and minimum operable frequency on each transit route within the bus network. It depends, typically, on the available fleet size and transit demand for each route. Lastly, (5.7) puts a constraint on the maximum number of routes or network size. This is generally determined by transit authorities who stipulate the number of routes they can provide. In practice, this constraint depends on the available financial resources, which the authorities can invest in operating the network.

5.3.2 Network generation

This is the first stage of the SBO model. It deals with creating a pool of feasible transit networks, from which the first population of solutions will be initialised in the model's optimisation framework. An ad-hoc heuristic algorithm was developed for the network generation exercise. Its inputs include 1) an existing transit network and its constituent routes, 2) the network size parameter (the number of routes), and 3) feasibility criteria for route length r_{len}, route directness r_{dir} (minimum deviation from the shortest path), and route overlap $r_{overlap}$ (maximum coincidence between the links of a route and that of the shortest path). These parameters will be used to define the feasibility conditions for acceptable routes. The network generation heuristic is developed using the Java programming language (Arnold and Gosling, 2000), *JGrapht* (Michail et al., 2019)—an open-source graph creation and manipulating the programme, and the Extensible Markup Language (XML) (Bray et al., 2006). The existing transit network data is presented as one of the outputs of a General Transit Feed Specification (GTFS) feed for a current public transportation service. This involves extracting and reformatting the transit network and routes from the available GTFS data. The network is then converted into a *GraphML* file (Brandes et al., 2002), which is a unique XML format for graphs. The conversion makes it possible to read the network as a graph with its nodes, links and their attributes. Subsequently, the graph can be manipulated with the JGrapht tool and graph theory operations. As part of the reformatting activity, the Origin Destination (OD) stops for existing network routes are extracted and used in the NGA. The network generation procedure is discussed next.

Network generation algorithm The steps taken generate the feasible candidate networks for the SBTNOM's NGA are:

1. **Read in OD pair data**

   The algorithm starts by reading in the OD pairs that have been extracted from the existing network routes.

2. **Generate multiple paths between each OD pair**

   Subsequently, the k-shortest paths algorithm by Yen (1971) is used to create a user-specified number of paths for all the OD or node pairs. This way, multiple routes can be enumerated between each OD pair. The k-shortest path algorithm, typically, generates multiple paths in increasing order of magnitude relative to a weighted cost factor. In this work, the path length in kilometres for each route is used as the cost factor. Therefore, if x paths are generated between an OD pair, the first path corresponds to the Dijkstra Shortest Path (SP) (Johnson, 1973), and its length is equal to the beeline distance between the node pairs. The created paths, which will hereafter be referred to as *alternate* paths, are usually longer than the SP in increasing order of magnitude.

3. **Check route length feasibility conditions for all routes**

   At this stage, the route length r_{len} feasibility is checked for both the shortest path and alternate paths. This is done to verify that a maximum and minimum route length condition is satisfied.

4. **Check other feasibility conditions on the alternate paths**

After satisfying the route length feasibility, other checks are then carried out on only the alternate paths. The first one verifies the directness r_{dir} of the route. It checks that an alternate path does not deviate excessively from the geometry of the shortest path.

Check for route directness This is important because users consider route deviations unappealing; hence, the deviation needs to be minimum. However, it is sometimes necessary for a route to deviate to adjoining areas where a major transit route does not run. This may help to cover the demand in those areas. A factor of 1.2 is used in this work.

Check for route overlap The second feasibility condition is for route overlap $r_{overlap}$. It checks whether there is a similarity between the links of the SP and the alternate path. A minimum value of 0.5 has been used in this work. This implies that each satisfactory alternate route must contain at least half of the SP's constituent links. Walker (2011) Recommends that overlap should be kept minimal, because they may increase the total cost of operating a network without necessarily improving the service frequency.

Check if the route exists already Lastly, a final check is made to ensure that the alternate path does not currently exist in the list of stored routes.

5. **Save the feasible routes for the current OD pair**

If all the above-stated conditions are met, the alternate path is saved as a candidate route in a list created for the specific OD pair. This process is then repeated for all the OD pairs, with each OD pair having its own unique list wherein the routes generated for that OD are saved.

6. **Perform stratified sampling of routes in all saved list of routes**

After generating the feasible routes, candidate network solutions are created by first using a stratified sampling technique to select routes from each OD, and then combining them into a network. In stratified sampling, a population is divided into various sub-populations. Individuals are then selected from each group or strata to make up a random sample. See further details of stratified sampling in Dorofeev and Grant (2006). Drawing from this sampling technique, in this work, the list of routes generated for each OD pair is considered a *stratum*. The sampling is then achieved by randomly choosing one route from each stratum, and combining the selected routes into networks.

This process ensures that the order of the existing network ODs pairs is retained after sampling. Through this process, a pool of 1500 feasible networks is generated. The first population is then initialised in the NSGA-II from this pool of feasible networks. In cases where retaining the order of the routes is not paramount, the feasible networks generated for all the OD pairs can be placed in a single pool. Other sampling techniques, like *random sampling*, may then be used to generate the required number of networks.

7. **Convert the sampled routes to a network transit schedule input file**

   The final step in the route generation process involves converting the candidate route networks to MATSim `transitSchedule.xml` files which is the appropriate input format for the optimisation algorithm. However, for the NSGA-II to operate on the solutions, a unique encoding will be defined. Details of this will be seen when the NSA is discussed.

5.3.3   Network evaluation procedure

This step of the model involves using the MATSim to evaluate the generated network solutions. It requires setting up a MATSim public transit scenario for the problem, which is called by the SBTNDM during the evaluation process. Inputs for the scenario include, 1) the initialised population of network alternatives, 2) a synthetic population of agents and their travel demand or (24-hour activity plan), which is created from the AFC data, 3) an initial schedule of transit operations on the routes of the network, comprising a timetable with its detailed fleet schedule and vehicle departures, and 4) a fleet of transit vehicles that will operate the schedules. The analysis is done with parameter values from the scenario. MATSim is called each time a new solution is to be evaluated.

Before evaluating a new solution, the subsisting transit schedule data file is overwritten since it would have been altered during the NSGA-II reproduction. The MATSim simulation process then begins with executing the users' initial demand and optimising them. At the end of the simulation, the resulting events file are analysed. This analysis involves evaluating the objective functions in (5.2) and (5.3), respectively, with parameter values obtained from the events file. A score or objective function value is obtained from the analysis. The score is assigned to the current network solution, which is then returned to the optimisation module for further processing. The MATSim scenario used in the **book** was parallelised. Therefore, the final evaluation score is realised by aggregating the scores of individual instances of the simulation. The parallel implementation of the simulation is discussed next.

Parallelising MATSim One way to account for the randomness associated with stochastic processes is to simulate the process multiple times and use the mean result of the different simulation runs. In this work, the simulation experiment involves running multiple instances of MATSim in each evaluation of the candidate network solutions. This implies that to satisfactorily describe the stochastic behaviour of passengers on the transit network, multiple runs of the simulation are required in each iteration of the optimisation process. MATSim possesses multi-threading capabilities, which means that it can run in parallel when extensive simulations or a high number of iterations is required. The parallelisation is achieved by setting MATSim's `numberOfThreads` feature in the `global` module within the configuration file. A MATSim scenario comprising of multiple runs of the simulation is set up and configured to run in parallel. Internally, each *simulation* or *run* comprises of a user specified number of MATSim iterations. For this work, it was determined experimentally in section 3.3.2 that each simulation converges after 80 iterations. It should be noted that the iterations operate sequentially and not in parallel. This is because, in the simulation, each new iteration uses the results of the previous one as input. In essence, succeeding iterations *learn* from preceding ones till equilibrium is achieved in the simulation. However, since multiple *runs* or *simulations* are required in this case, they can be set up in parallel. Each parallel MATSim simulation is executed in its own Java Virtual Machine (JVM) because each one needs to use a unique Pseudo Random Number Generator (PRNG) (Matsumoto and Nishimura, 1998; Rahimov et al., 2011). In the end, the various results are averaged and used. The collection of multiple *runs* will be referred to as an *ensemble of runs*.

It is this ensemble of runs that is referred to as one *evaluation* of each candidate network. In this **book**, 30 runs of the simulation are used in each ensemble. This means that to evaluate a network once, 30 parallel runs of MATSim is executed. Therefore, in an instance where a sample problem has a population of 10 networks and the termination criteria is set to 10 generations the total Number of Function Evaluations (NFEs) required is 100. This is obtained from (population size x number of generations). However, since each *evaluation* comprises of 30 MATSim runs, the total number of MATSim runs will be (population size x number of generations x ensemble size) i.e. 10 x 10 x 30 = 3000. An image depicting the MATSim *ensemble* may be seen in Figure 5.2.

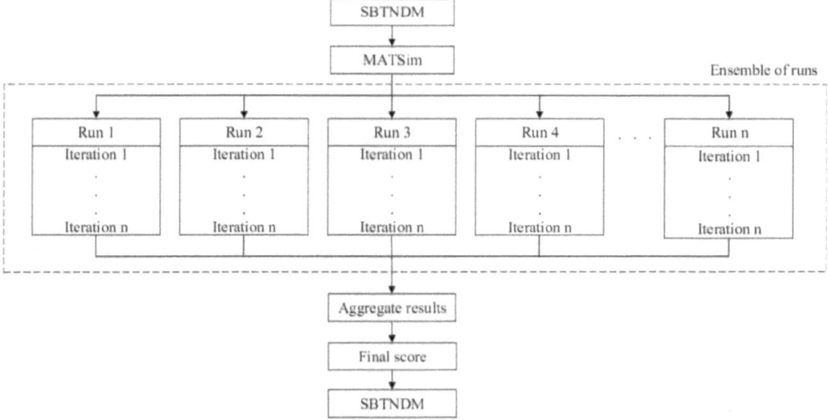

Figure 5.2: The parallel implementation of MATSim used in the SBTNDM.

5.3.4 Network optimisation

This stage of the model describes how the network optimisation progresses and how a Pareto set of transit network solutions is realised. The main inputs used here are the outputs of the NGA and NEP, respectively, namely the feasible candidate solutions and objective function scores from simulating each solution with MATSim. This implies that at different stages of its operation, the procedure will call the NGA and NEP sub-routines. The generated network routes are converted to MATSim transit schedule files, which contain both the transit routes and their schedules. Hence, the *solutions* or chromosomes in this section, refers to MATSim `transitSchedule.xml` files. The format of these files is XML, which can also be referred to as the solution's phenotype. An important step in evolutionary algorithms is to encode the gene-level operation of each solution. To this end, the `transitSchedule.xml` file, which is initially in XML format is converted to a Java Script Object Notation (JSON) data format. This facilitates the efficient and straightforward manipulation of the transit schedule with the genetic operators (selection, crossover and mutation) during the reproduction process. However, this encoding scheme then makes it necessary to customise the NSGA-II operators to enable them manipulate the JSON format. As stated previously, the major advantage of this approach is that it accommodates the encoding of each network with a detailed operational schedule. It also facilitates the simultaneous handling of the route network design and frequency setting sub-problems of the TNDP.

Network search algorithm The optimisation process starts with randomly selecting 100 out of the pool of 1500 feasible solutions created in the NGA. These are then initialised in the NSGA-II. The initial population is evaluated with the NEP. The NSGA-II's crowding comparison operator is then used to rank the solutions. Subsequently, pairs of networks are drawn randomly to serve as parents for the reproduction process. After that, the selected parents are encoded as JSON files. A custom version of the *single point crossover* and *polynomial mutation* are then used to perform the reproduction of offspring. These customised operators are capable of manipulating the JSON file format. First crossover is done by swapping a given number of corresponding network routes between both parents at a specified crossover point. Hence, offspring network solutions which have similarities with the parent networks but are different in configuration are created. After this, mutation is done by replacing one route in each newly created offspring network with a route that is randomly selected from the original pool of networks created by the NGA. The crossover and mutation operators are controlled by probabilities, which are set to 0.75 and 0.25 respectively. This process continues repeatedly, till the appropriate population size of offspring is created. After creating this new population of offspring network solutions, they are merged with the parent population. The combined population is then evaluated and ranked again. The better performing half of the population is then selected again to restart the reproduction. This process continues iteratively, until the termination criterion (number of generations) is reached. This is set to ensure the algorithm will stop once the criterion is satisfied. Lastly, the set of solutions, obtained in the final generation, are decoded by converting them from the JSON format back to the MATSim transit schedule files in XML format. This solution set is the Pareto solution set for the transit network design problem at hand and may be subjected to further analysis where necessary. A summary of the main components of the SBTNDM in terms of their input data, parameters and their output can be seen in Appendix B.1.

5.4 Summary

This chapter focused on giving a full description of the SBTNOM and its procedures. First, the data model, modelling assumptions and a mathematical model for the SBTNDM are discussed. Following this is a description of the solution framework, which looks at the three main steps taken to integrate MATSim and NSGA-II. The steps are the NGA which generates candidate network solutions, the NEP which evaluates the solution, and the NSA that searches for the optimal network solution. A significant part of this description focused on how new solutions to the problem are evolved within the solution framework, resulting from the feedback loop between the simulation and optimisation components of the SBTNDM. The next chapter discusses the computational testing of the SBTNOM.

Chapter 6

Model testing

This chapter discusses various tests carried out to determine how effective the Simulation-based Transit Network Optimisation Model (SBTNOM) is in designing transit networks. The tests mainly focus on the computational performance of the model. The three main tests discussed here are: 1) repeatability tests, 2) performance or quality evaluation, and 3) a comparison or benchmarking of the SBTNOM when implemented with other Multi-objective Evolutionary Algorithms (MOEAs) that are different from the Non-dominated Sorting Genetic Algorithm (NSGA-II).

The repeatability tests are done to determine the level of variability in the results produced by the Simulation-based Transit Network Design Model (SBTNDM). In terms of quality evaluation, five performance indicators will be used to measure the quality of the SBTNOM's solutions. The indicators are 1) the hypervolume indicator, 2) generational distance indicator, 3) epsilon additive indicator, 4) the maximum Pareto error indicator, and 5) spacing indicator (Coello et al., 2007; Michalewicz, 1992; Talbi, 2009a). In this particular test, the model's attributes like the spread and convergence of its solutions in the Pareto front are reported.

The final computational test will focus on comparing the SBTNOM with other MOEAs. This technique was previously used by Laumanns et al. (2005) and Chase et al. (2009). The indicators mentioned above are used once again to evaluate how the different MOEAs perform. The SBTNOM is a versatile tool which allows integration with other optimisation algorithms using minimal effort. This feature makes it possible to observe and compare how different algorithms will perform on a specific Transit Network Design Problem (TNDP) when integrated with the SBTNOM. The computational tests discussed in the chapter were carried out with the *MOEAFramework*. This is a computational laboratory built by Hadka (2017) for solving various Multi-objective Optimisation Problems (MOPs). The tests reported in this chapter were conducted with a Dell Linux High Performance Computing (HPC) cluster with a total of 1368 nodes and 32832 cores. 16 nodes having 125gb and 24 cores each were dedicated for testing the SBTNOM. The Centre for High Performance Computing (CHPC), South Africa provided the resources. These tests are reported next.

6.1   Model repeatability

This deals with the variation in the results obtained from a simulation or experiment under constant conditions (Bartlett and Frost, 2008). In the context of this **book**, *repeatability* checks the amount of variation between the results of multiple runs of the SBTNOM, assuming the measurements are taken by the same researcher and with similar input parameters. In the SBTNOM, each evaluation of the model's solution, comprise multiple parallel runs of the Multi-Agent Transport Simulation (MATSim) simulation referred to in section 5.3.2 as an *ensemble of runs*. The simulation is run several times because of the uncertain nature of dynamic and stochastic simulations like MATSim. Each run of the simulation is therefore performed with a different *random seed* to obtain different simulation results. In computational science, random seeds are used to generate a series of pseudo-random numbers that can replicate the state of an experiment or simulation (Gosavi, 2015c). It implies that if all input parameters are kept constant, a simulation's results would be the same if set to run with a given seed, and different if the random-seed is different. It is also noted that with a higher number of simulation runs in the ensemble of runs; it is possible that better test results will be obtained. However, the number of runs used in this work is limited by the availability of computational resources. The repeatability test is described next.

6.1.1 Confidence intervals

A 95% Confidence Interval (CI) test is used to measure and report the repeatability of the SBTNDM. In statistics, CI indicates a range of values or intervals, within which the true value of a population parameter is likely to be found. The interval is commonly referred to as *confidence limits*, which denotes the upper and lower limits of the CI. It is obtained by subtracting and adding the CI to the sample parameter such as a *sample mean*. The CI is normally discussed relative to a Confidence Level (CL), which is the probability that the CI contains that true parameter value. In this work the CI is presented within a CL of 95%. In terms of the configuration used, the test comprises 25 experiments. Each experiment has a *sample size* of 30 networks and 50 *generations*. Hence, the total Number of Function Evaluations (NFEs) is 1500. Lastly, each *evaluation* or *ensemble of runs* comprise of 30 MATSim simulations running in parallel. The results of the repeatability tests are discussed relative to the user and operator cost objective functions. In Table 6.1 the results obtained from the test in terms of the user cost objective in hours is presented. The tables show the CI values for the individual experiments as well as their Relative Standard Error (RSE). The RSE measures the variability of a given *quantity* or *statistic* within a sample. This implies that it checks how the *observed sample parameter* such as a *sample mean* changes as the same experiment is repeated any given number of times. The RSE is obtained by expressing the Standard Error (SE) as a percentage of the observed sampled statistic (mean in this case). For instance, if the SE of an experiment is 5 and the sample mean is 75, the RSE for that experiment is obtained by $(5 \div 75) \times 100$, which gives 6.67%. Experiments with a high RSE are commonly perceived to have a high sampling error. In the literature, there is not a consensus of what constitutes an acceptable threshold for RSE values. This is because the context, variability of the input data, and parameters of all experiments are unique. However, it is widely agreed that lower values of the RSE are more acceptable and a sign that an experiment is repeatable. Consequently, smaller values of RSE indicate a low sampling variation around the mean or less spread. For this reason, smaller values are considered more acceptable. Table 6.2 presents descriptive statistics for the experimental sample, while Figure 6.1 shows a plot of the CI.

Table 6.1: Results of the experimental runs from the SBTNOM in terms of user cost objective.

Exp No	Mean μ	Std Err	RSE %	CI	μ - CI	μ + CI
1	683.535	2.487	0.364	4.935	678.600	688.470
2	688.808	2.592	0.376	5.143	683.665	693.951
3	692.691	2.710	0.391	5.377	687.314	698.068
4	691.017	2.646	0.383	5.250	685.767	696.266
5	687.047	2.451	0.357	4.862	682.185	691.910
6	690.969	2.595	0.376	5.150	685.819	696.118
7	691.560	2.461	0.356	4.883	686.676	696.443
8	690.848	2.499	0.362	4.959	685.889	695.807
9	683.485	2.290	0.335	4.545	678.940	688.029
10	691.615	2.372	0.343	4.706	686.910	696.321
11	684.925	2.413	0.352	4.789	680.136	689.713
12	689.113	2.419	0.351	4.801	684.313	693.914
13	687.752	2.576	0.375	5.112	682.640	692.863
14	695.054	2.343	0.337	4.649	690.405	699.703
15	690.923	2.525	0.365	5.009	685.913	695.932
16	691.886	2.411	0.348	4.784	687.102	696.670
17	686.473	2.571	0.375	5.102	681.371	691.575
18	692.170	2.450	0.354	4.862	687.308	697.032
19	687.907	2.488	0.362	4.937	682.970	692.844
20	689.950	2.635	0.382	5.228	684.722	695.178
21	687.253	2.300	0.335	4.563	682.690	691.816
22	690.992	2.513	0.364	4.987	686.005	695.979
23	690.438	2.347	0.340	4.656	685.782	695.094
24	684.406	2.434	0.356	4.830	679.576	689.235
25	686.401	2.286	0.330	4.536	681.865	690.937

Table 6.2: Summary statistics for the 25 experiments.

Parameter	Value
Mean	689.089
Standard Error	0.612
Relative standard error (%)	0.089
Median	689.950
Standard Deviation	3.059
Minimum	683.485
Maximum	695.054
Count	25
Confidence Level(95.0%)	1.263

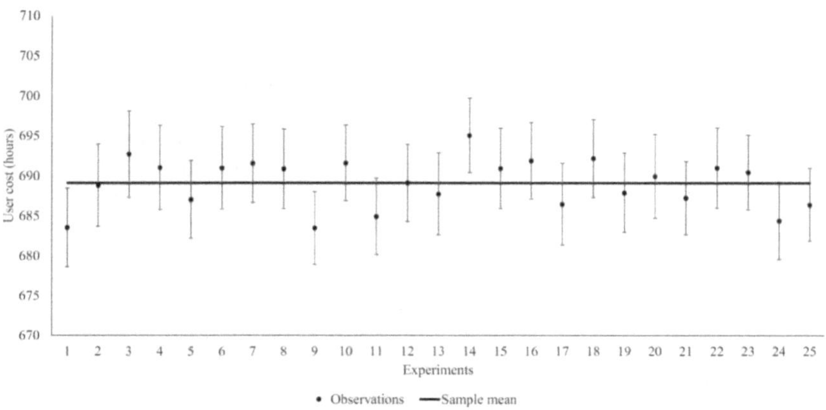

Figure 6.1: Plot of the confidence intervals for the 25 experiments.

Figure 6.1 shows that of 25 the experimental results produced by the model, 22 contain or overlap with the mean parameter. Therefore, for the distribution of 25 Transit Network Design (TND) experiments performed with the SBTNOM and reported for the user cost objective, the 95%CI is (687.826, 690.351). This means that for experiments conducted with the SBTNOM using different samples, 95 times out of 100, the *population parameter* would lie within the estimated CI values.

In terms of the operator objective in monetary units, Table 6.3 show the results of the experiments with their CI and RSE. Table 6.4 gives a descriptive statistics of all 25 experiments, while the plot for the CI may be seen in Figure 6.2.

The results depicted below shows that 23 contain the mean parameter. Therefore, for the distribution of 25 TND experiments performed with the SBTNOM and reported for the operator cost objective, the 95%CI is (26070.774, 26136.273). This means that for experiments conducted with the SBTNOM using different samples, 95 times out of 100, the *population parameter* would lie within the estimated CI values. In light of these results, it can, therefore, be inferred that the SBTNDM's results are repeatable. Having satisfied this important criterion, in the next section, a different test, which evaluates the quality of the SBTNOM's solutions, is discussed.

Table 6.3: Results of the experimental runs from the SBTNOM in terms of operator cost objective.

Exp No	Mean μ	Std Err	RSE %	CI	μ - CI	μ + CI
1	26229.190	80.651	0.307	160.030	26069.161	26389.220
2	26060.189	67.574	0.259	134.081	25926.108	26194.270
3	26047.513	73.925	0.284	146.684	25900.830	26194.197
4	26072.750	70.163	0.269	139.218	25933.532	26211.969
5	26214.419	75.781	0.289	150.047	26064.373	26364.466
6	25950.890	72.368	0.279	143.594	25807.297	26094.484
7	26067.030	81.059	0.311	160.839	25906.191	26227.868
8	26078.553	83.141	0.319	164.969	25913.584	26243.522
9	26150.866	61.065	0.234	121.166	26029.699	26272.032
10	26092.686	83.620	0.320	165.921	25926.765	26258.607
11	26291.697	79.289	0.302	157.326	26134.371	26449.023
12	26153.453	83.923	0.321	166.522	25986.931	26319.976
13	26083.808	80.474	0.309	159.677	25924.131	26243.486
14	26054.420	86.578	0.332	171.790	25882.630	26226.211
15	25990.803	72.326	0.278	143.511	25847.292	26134.315
16	26095.785	77.729	0.298	154.230	25941.555	26250.015
17	26050.819	64.519	0.248	128.019	25922.799	26178.838
18	26102.833	95.895	0.367	190.277	25912.557	26293.110
19	26136.534	69.885	0.267	138.667	25997.867	26275.201
20	26125.091	88.632	0.339	175.866	25949.226	26300.957
21	26046.195	86.269	0.331	171.177	25875.018	26217.373
22	26046.997	85.502	0.328	169.654	25877.343	26216.651
23	26042.450	78.230	0.300	155.225	25887.226	26197.675
24	26234.554	75.330	0.287	149.470	26085.084	26384.024
25	26168.560	79.997	0.306	158.732	26009.828	26327.291

Table 6.4: Summary statistics for the 25 experiments.

Parameter	Value
Mean	26103.524
Standard error	15.868
Relative standard error (%)	0.061
Median	26083.808
Standard deviation	79.339
Minimum	25950.890
Maximum	26291.697
Count	25
Confidence level(95.0%)	32.750

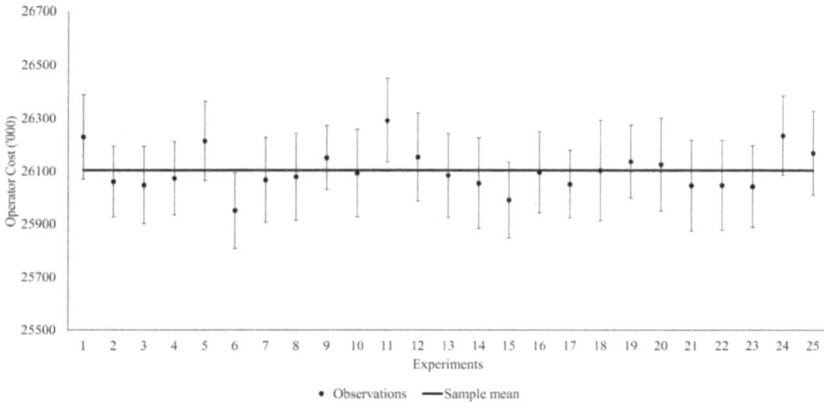

Figure 6.2: Plot of the confidence interval for the 25 experiments in terms of operator cost.

6.2 Quality evaluation

When solving a MOP, the goal is to find a Pareto set of solutions if it exists. However, the difficulty of problems like the TNDP make researchers opt for a locally optimal solution set which is usually the best approximation of the Pareto optimal set and acceptable given the limited time resources. Therefore, the quality of MOEA solutions is measured based on the proximity of a locally optimal solution set or *approximation set* to the global Pareto optimal set or *reference set* if the latter is known or available. By checking factors like convergence or the spread of solutions across the Pareto front, it is then possible to measure the quality of a MOEA's results. In this section, one objective is to evaluate the quality of the SBTNDM's results with different MOP performance indicators. Another goal is to observe if the SBTNDM behaves as reported in the literature relative to these indicators. Furthermore, the tests are conducted with a population size of 30 networks and runs for 50 generations respectively. Therefore, the tests will reveal the generation at which convergence will occur. This is useful because, for the above-stated population size, the number of generations at which good trade-off network solutions may be obtained is not known. Therefore the results of these tests will help to identify the

108

optimal number of generations required for acceptable results to be obtained. These tests can be conducted to determine the optimal values of other parameters for the model. However, a detailed sensitivity analysis of the SBTNOM is outside the scope of this book. The results are collected and recorded as the model runs then analysed and plotted at the end of the model's execution.

6.2.1 Hypervolume

This indicator is one of the most common metrics that is used to evaluate the quality of a MOEA's solution. It measures the size of a MOP's search space that is dominated by the problem's approximation set (Bringmann and Friedrich, 2013). The approximation set is the set of sub-optimal solutions or a local Pareto front which is an approximation of the globally Pareto front. The indicator is calculated relative to a reference point known as the *nadir* point (Hadka, 2017). This is usually the worst case objective value for each objective function. Some advantages of the hypervolume indicator are that 1) it is easily adapted to problems with many objectives, 2) it is a measure of both convergence and diversity in a MOEA, and 3) it does not require prior knowledge of the Pareto front to guide the search for a solution that approximates the former. The limitation of this indicator is that it is computationally expensive.

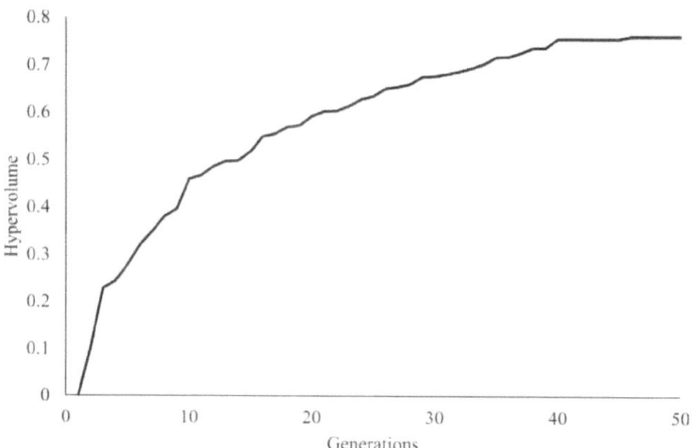

Figure 6.3: Plot of the Hypervolume indicator

In terms of its behaviour, a higher value of the hypervolume indicates a better solution or approximation set, because it dominates a greater portion of the search space. Figure 6.3 shows a plot of the indicator after 50 generations of the SBTNDM. The figure shows that the value of the indicator steadily increases as the algorithm's generations increase. This implies that the SBTNOM's solutions improve in successive generations which matches the known behaviour of the hypervolume indicator. The results also show that the indicator converges close to 50 generations, hence, the number of generations to get optimal network solutions with the SBTNDM is 50.

6.2.2 Generational distance

The Generational Distance (GD) indicator is obtained by measuring the average distance between each solution in the approximation set and the nearest one in a MOP's reference set. Since the indicator measures proximity to the reference set, it is a good indicator for measuring convergence in MOEAs (Liu et al., 2019). Furthermore, smaller values of the indicator are considered better. When the approximated set is a subset of the reference set, the GD is equal to zero. The GD alone may not be sufficient to evaluate the quality of a MOEA. A plot of GD against the number of generations is shown in Figure 6.4.

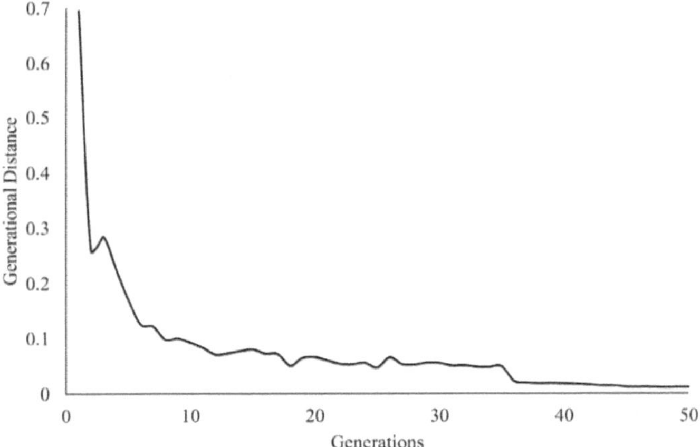

Figure 6.4: Plot of the Generational distance indicator.

110

If the approximation set contains a single solution that is too close to the reference set relative to other solutions in the set, the GD measurements may be unrealistically low. For this reason, it is often combined with other quality or performance indicators. The results show a convergence of indicator values after 45 generations. The behaviour of this indicator observed in the figure is in line with the expected behaviour of the GD that was previously discussed.

6.2.3   Epsilon indicator

The epsilon or additive indicator measures the smallest distance needed to translate an approximation set to dominate the reference set completely. For this indicator, lower values of the distance needed to cover the reference set, indicate better proximity and diversity in the approximation set. However, if a region exists in the reference set which is poorly approximated by the approximation set, more considerable distances would be required. According to Hadka and Reed (2013), the additive indicator measures the consistency of an approximation set. Therefore, to achieve the best results, the approximation set should not contain large gaps or regions of poor approximation. In figure 6.5, the values of the indicator are plotted against the generation parameter of the SBTNOM. The plot shows the convergence of the indicator's values occurs after 45 generations.

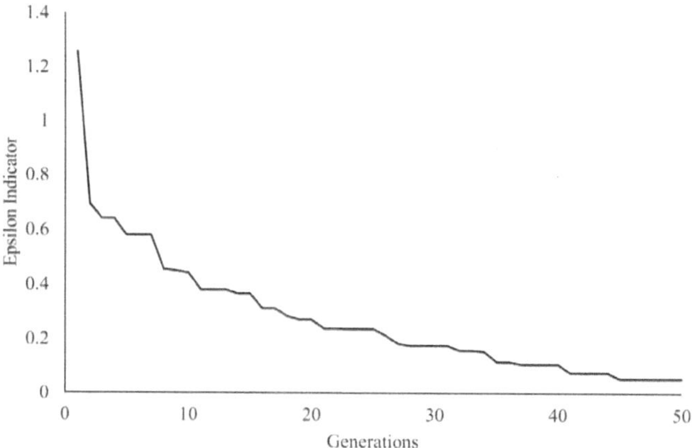

Figure 6.5: Plot of the Epsilon indicator.

6.2.4 Spacing

The spacing indicator measures the *spread*, or uniformity of the spaces between neighbouring solutions in an approximation set (Coello et al., 2007). The solutions in a well-spaced approximation set are usually not densely clustered in a local vicinity of the Pareto front. The spacing indicator does not need a reference set to be calculated (Hadka, 2017). Sometimes this leads to the approximation registering a good spacing while having a poor closeness to the reference set. It is therefore recommended that the spacing indicator be used with other performance metrics that measure proximity. In Figure 6.6, the values of the indicator are plotted against generation parameter of the SBTNOM.

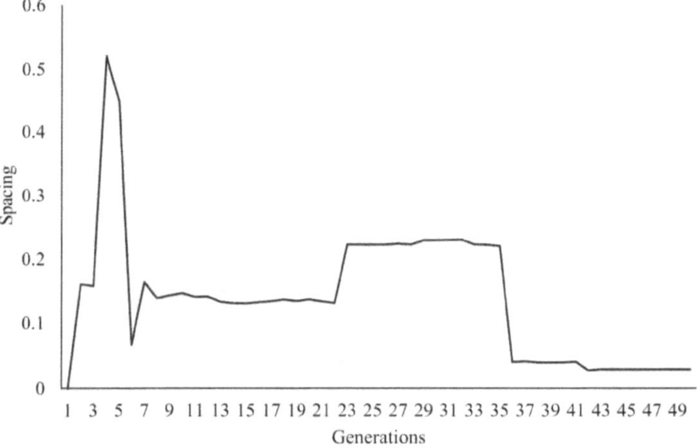

Figure 6.6: Plot of the Spacing indicator.

The spacing metric shows convergence after 43 generations.It can, therefore, be concluded that the solutions obtained after 43 generations have a more even spread within the search space.

6.2.5 Maximum pareto front error

This indicator evaluates how well a set of solutions compares to another. It measures the largest minimum distance between each solution in the approximated set and a corresponding nearest solution in the reference set (Yen and He, 2014). Normally, a well-spaced approximation set does not possess dense clusters of solutions that are separated by large empty spaces. This indicator too, is estimated without a reference set. Consequently, an approximation set may register good spacing even though its proximity to the reference set is bad. To this end, it is recommended that the spacing indicator be used along with other performance metrics. In Figure 6.7, the values of the indicator are plotted against generation parameter of the SBTNOM.

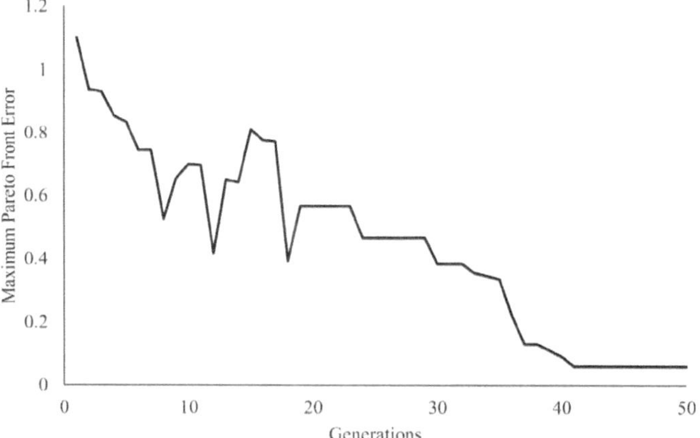

Figure 6.7: Plot of the Maximum Pareto Front Error indicator.

The figure shows a decrease of the indicator values as the generation parameter increase. This shows an improvement of the obtained solutions in successive generations of the SBTNOM. This is aligned with the accepted behaviour of the spacing indicator. The values converge after 40 generations implying that, for the indicator, the SBTNOM will get good solutions after 40 generations.

6.3 Pareto front visualisation

A Pareto front or its approximation set is the graphical representation of non-dominated solutions in the search space of a MOP. Typically the Pareto front should give the Decision Maker (DM) complete information about the feasible optimal solutions in the search space. Convergence and diversity are the two major characteristics of a Pareto front. Convergence shows the proximity of an approximation set to the reference set or true Pareto set. In problem cases like in this **book** where the true Pareto front is not known, the previously discussed indicators are used to show the convergence of the solutions. In terms of diversity, it is expected that the solutions in the Pareto front should be spread out uniformly across the front. This implies that there should not be cluster(s) of solutions or voids in the Pareto front. A lack of either convergence or diversity in the obtained non-dominated solution set indicates that the latter is not the best approximation of the true Pareto front. Hence, they would not convey all the useful information a decision-maker needs to decide on the obtained solutions. To this end, visualising the Pareto front is an essential tool for informing the DM about the obtained solutions in the MOPs (Agrawal et al., 2004; Lotov and Miettinen, 2008). Furthermore, visualisation is important as it can help us transform symbolic data to geometric information which aids humans in forming mental pictures of the symbolic data.

According to McQuaid et al. (1999), for visualisation to be effective, it has to be easily understandable and possess all the relevant information required to grasp the image quickly. If a visualisation technique satisfies these requirements, the image of the Pareto front can help the DM make a preliminary analysis of the solutions. That way, they can identify the region of the front where the good trade-off solutions are likely to be found. Those solutions may then be isolated for further objective analysis to identify which one best addresses the DM's needs. In this **book**, a scatter plot technique is used to visualise the SBTNOM's Pareto front. Figure 6.8 show the Pareto front obtained after 15, 30 and 50 generations respectively.

An observation of the plot reveals that the solutions obtained after 15 generations are spreads out further across the search space (along the *x-axis*) than the front obtained after 30 generations. Similarly, the Pareto front after 30 generations is spread out further than that obtained after 50 generations. This implies an in-depth exploration of the search space has been done, which is indicative of a good diversity among the obtained solutions. As better solutions are found in each generation, the algorithm need not search parts of the search space where these results have previously been found. Furthermore, the fact that there are fewer solutions as the

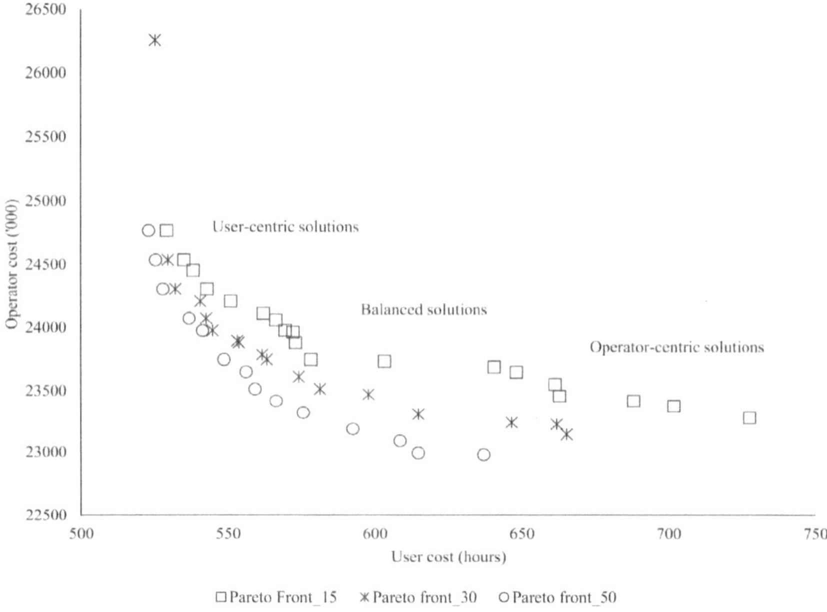

Figure 6.8: Pareto front after 15, 30 and 50 generations of the SBTNOM

algorithm runs progress shows that worse solutions are being eliminated. This is a sign of convergence in the results. Lastly, it can also be observed that the Pareto front after 50 generations has lower objective function values for both the user and operator cost functions. This is in line with the expected behaviour of minimisation problems.

6.4 Benchmarking the SBTNOM

In this section the SBTNOM is compared with other popular MOEAs in the literature. Recall that the model was implemented with the NSGA-II. The objective of this comparison is to determine if the SBTNOM performs better with other algorithms. The comparison is made relative to the previously discussed indicators, and the algorithms under consideration are the Strength Pareto Evolutionary Algorithm (SPEA2), Non-dominated Sorting Genetic Algorithm (NSGA-III) and Indicator

Based Evolutionary Algorithm (IBEA). These algorithms are described next followed by a discussion of the results obtained from the benchmarking exercise.

6.4.1 Strength pareto evolutionary algorithm (SPEA2)

The SPEA2 was developed by Zitzler et al. (2002). Its predecessor is the Strength Pareto Evolutionary Algorithm (SPEA) Zitzler et al. (2000). It incorporates elitism through the use of an external archive which contains the previously obtained non-dominated solutions. At each generation, non-dominated individuals are copied to the external non-dominated set. For each individual in this external set, a strength value is computed. This strength is similar to the ranking value of the Multi-objective Genetic Algorithm (MOGA) by Fonseca and Fleming (1993). In SPEA2, the fitness of each member of the current population is computed. This is done, according to the strengths of all external non-dominated solutions that dominate it. The fitness assignment process considers both the convergence and diversity of approximated Pareto solution sets. In this assignment strategy, the number of individuals that dominate a solution and those dominated by it is considered simultaneously. Pareto dominance are then used to ensure that the solutions are adequately distributed along the Pareto front. Since the archived solutions play a role in the selection process of SPEA2, as its size grows too large, it might reduce the selection pressure, thus slowing down the search. Due to this situation, the authors decided to adopt a clustering technique that prunes the contents of the external non-dominated set so that its size remains below a certain threshold. According to López and Coello Coello (2009) some differences between the SPEA2 and SPEA are: 1) the former uses a nearest neighbour density estimation technique which guides the search more efficiently, 2) SPEA2 also has an enhanced archive truncation method which ensures the preservation of boundary solutions.

6.4.2 Non dominated sorted algorithm (NSGA-III)

The NSGA-III is a reference-point based version of the NSGA-II. It highlights population members that are non-dominated and near a set of defined reference points within the search space. The first step is to define the set of reference points. After that, the current parent population is used to create an offspring population by the action of genetic operators. The parents and offspring populations are then combined. Next, the combined population is sorted into different levels of non-domination using the same dominance ranking method used in the NSGA-II. After ranking, solutions that will be used to create the next generation are selected. In the NSGA-II, solutions having the largest crowding distance values or widest diversity are selected. This crowding distance operation is not amenable to problems with four or more objectives (Deb and Jain, 2013). Hence, the selection mechanism in the NSGA-III is modified by introducing a set of reference points, populated on a normalised hyperplane of the solution space. The strategy guarantees that the diversity of the solutions can be maintained even in higher-order objective problems. It results in an even distribution of non-dominated solutions across the objective space. The preferred non-dominated solutions are, therefore, the ones closest to the defined reference points. When compared with the NSGA-II, the main limitation of the NSGA-III is its higher computational requirement (Jain and Deb, 2014).

6.4.3 Indicator based evolutionary algorithm (IBEA)

The main distinguishing attribute of indicator-based MOEAs is that their fitness assignment scheme is based on quality indicators like the hypervolume or additive epsilon indicator. This is in contrast to many other MOEAs that use Pareto dominance methods. The IBEA by Zitzler and Künzli (2004) is one of the most widely documented examples of the indicator-based MOEAs. It uses the hypervolume performance indicator as a means of ranking solutions. The algorithm works on the principle that a quality indicator highlights solutions with desirable qualities. An initial population is generated in the first generation. Two individuals are then selected from the current population with a binary tournament selection operator. Next, crossover and mutation are used as genetic operators to reproduce two offspring. The generated offspring are then combined with the parent population to form the next generation of individuals. After this, the new population is evaluated iteratively, and the worst-performing solutions removed from the lot in each iteration until the number of individuals for the next reproduction process is reached. The process is known as environmental reproduction (Phan and Suzuki, 2013). The main disadvantage of this indicator-based method is that measuring its performance becomes computationally expensive, especially with an increase in problem objectives.

6.4.4 Indicator results

In the Figures 6.9a through 6.10c, the results of performance indicators for the above mentioned MOEAs are presented graphically. The results are collected after 50 generations of the algorithms' run and plotted graphically. The plot in Figure 6.9a shows that the combination of SBTNOM and NSGA-II has a higher value for the hypervolume indicator than the other algorithms. This shows that the NSGA-II performs better, as a greater value of the hypervolume indicator, is indicative of better results. Furthermore, except for the case of the hypervolume where larger values are more desirable, smaller values are indicative of better results for other indicators. Hence looking at the results, it is clear that the NSGA-II has smaller values than the other algorithms. However, the SPEA2 is the second-best performing algorithm in at least three of the indicators, namely hypervolume, generational distance and epsilon indicator. This may suggest that the SPEA2 may be considered desirable to find a solution to this specific TNDP in the absence of the NSGA-II.

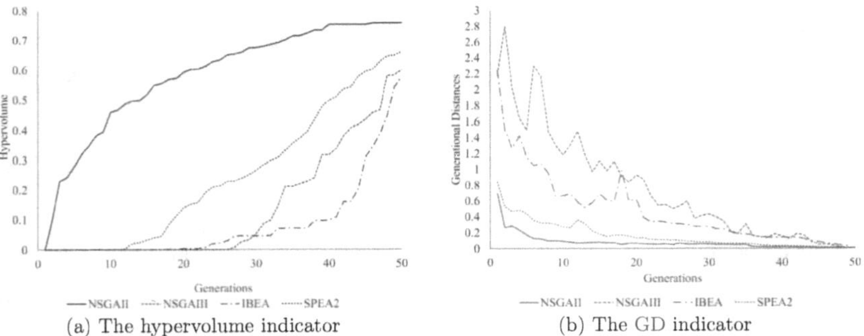

(a) The hypervolume indicator   (b) The GD indicator

Figure 6.9: Plot of all indicators after 50 generations for all four algorithms.

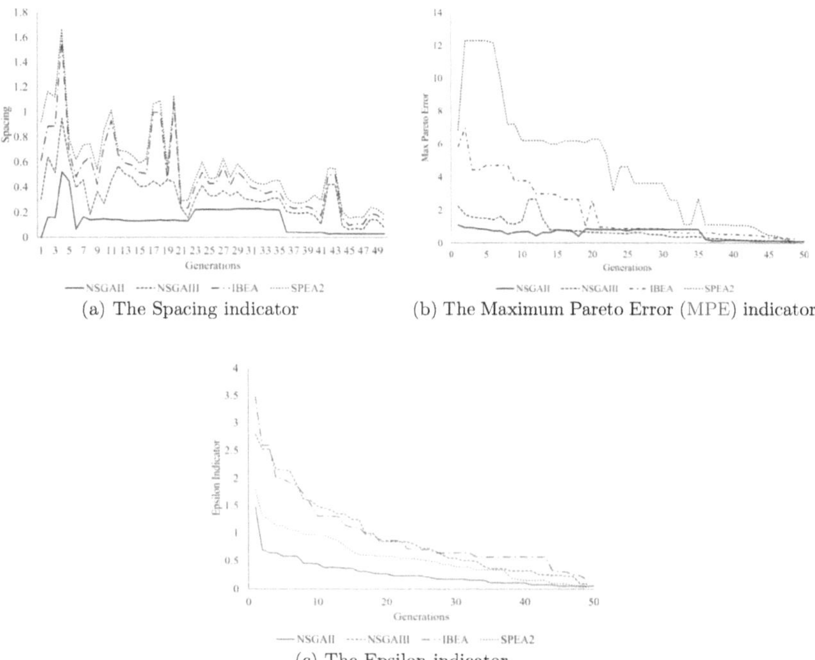

(a) The Spacing indicator   (b) The Maximum Pareto Error (MPE) indicator

(c) The Epsilon indicator

Figure 6.10: Plot of all indicators after 50 generations for all four algorithms cont'.

6.4.5 Pareto front results

In this section, plots of the Pareto front for all four algorithms that have been integrated with the SBTNOM may be seen in Figures 6.11a to 6.11d.

From the figures, the following may be deduced;

1. In terms of the user objective, the NSGA-II dominates the other algorithms except for the IBEA, i.e. in terms of the user cost objective the IBEA and NSGA-II are non-dominated relative to one another.

2. Though the NSGA-II does not dominate the IBEA in terms of the user objective, the spread of the IBEA's solutions in the search space is limited to a smaller neighbourhood when compared to the NSGA-II and SPEA2 .

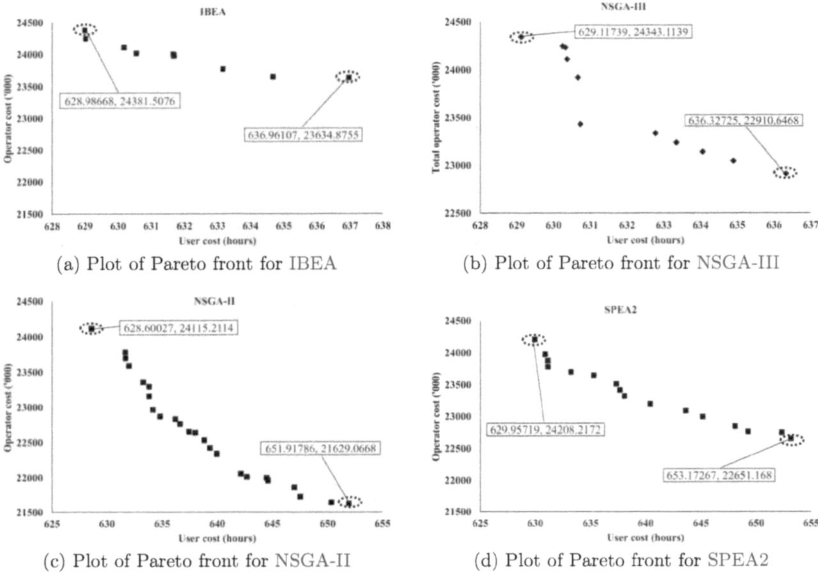

(a) Plot of Pareto front for IBEA

(b) Plot of Pareto front for NSGA-III

(c) Plot of Pareto front for NSGA-II

(d) Plot of Pareto front for SPEA2

Figure 6.11: Joint plot of Pareto front for the algorithms that were compared.

3. In terms of the second objective, which represents the operator's perspective, the NSGA-II outperforms the other algorithms by generating a solution with the least objective score.

4. Similar to the IBEA, the spread of the NSGA-III's solutions is very limited in comparison to the NSGA-II and SPEA2.

5. The IBEA and SPEA2 are non-dominated with respect to one another. This is because the IBEA dominates the SPEA2 relative to the user objective, while the SPEA2 dominates the IBEA in terms of the operator objective.

6. The NSGA-II and the SPEA2 are able to do an in-depth search of the solution space. They are also able to find a significantly higher number of solutions than the NSGA-III and the IBEA.

These observations coincide with the widely accepted definition of non-domination, which is that a solution is only considered as non-dominated if it performs better on

120

one objective and does not necessarily perform worse in all the other objectives. To this end, it may be inferred that the NSGA-II produced non-dominated solutions in comparison with the other algorithms and the context of this **book**.

In conclusion, the results discussed in this benchmarking exercise relative to the performance indicators and Pareto frontier of the four algorithms shows that the SBTNOM when integrated with NSGA-II outperforms the others, while the SPEA2 may be considered as the second-best among the lot. This confirms that the choice of the NSGA-II for this work was a good one.

6.5 Summary

In this chapter, a series of tests were conducted to determine the SBTNOM's computational performance. The first two tests focused on measuring the model's repeatability, and the quality of its results. The last test discussed a benchmarking exercise that is performed to compare the model's performance with other MOEAs. The results show that indeed the SBTNOM's are repeatable within the stated CI. Furthermore, concerning the performance or quality evaluation, the model's behaviour is as expected and in line with documented evidence in the literature. Lastly, the result of the benchmarking exercise shows that the SBTNOM performs better with the NSGA-II than with the other MOEAs. This justifies the choice of the NSGA-II as the optimisation algorithm that is used in this **book**.

Chapter 7

Model Application

In this chapter, the focus is on applying the Simulation-based Transit Network Optimisation Model (SBTNOM) which was developed in previous chapters to the design of a realistic and large-scale public transport network, namely the MyCiTi Bus Rapid Transit (BRT) in the City of Cape Town (CoCT). The application focuses on determining the network-related performance of the solutions obtained with the model. This will be done in terms of the respective objectives related to commuters and service operators on the MyCiTi system. Passenger ridership and Automated Fare Collection (AFC) data for the service has been obtained from the Transit Development Authority (TDA) for the exercise. Two applications will be considered. In the first application, the network solutions in the Pareto front obtained from the SBTNOM will be analysed. The analysis is done to determine the performance of the obtained network solutions. Public transport indicators will be used to evaluate the solutions, in particular, to see how realistic the model's results are. The passenger ridership data for a typical day is used in this case study. The second application involves designing networks that best satisfy the revealed demand for travel on the MyCiTi service over a twelve-month analysis period. Passenger ridership data for 12 different days representing the 85th percentile ridership level in each month of the period under review is used as input for this case. An elaboration of both applications will follow, but this is preceded by a brief description of the revealed MyCiTi AFC travel demand data.

7.1 The MyCiTi BRT revealed travel demand data

In terms of the revealed passenger ridership data on the MyCiTi BRT system, Figure 7.1a to 7.1d show the typical ridership statistics over a daily operational window. The plot reveals a profile of passenger transactions—hourly boardings, alightings and connections during operational hours. The highest ridership occurs at 7am during the morning peak period. The first boardings for the day occur at 4am, the first alighting transaction occurs at 5am while the service stops its operation at 10pm in the evening. On average 8000 riders used the service in the peak hour. Furthermore, the annual daily ridership on the system between 2015 and 2016 is just over 43000 daily. However, since the data was collected, there has been an increase in ridership on the service. A detail of the aggregated ridership data for the period under review is presented in Appendix C.

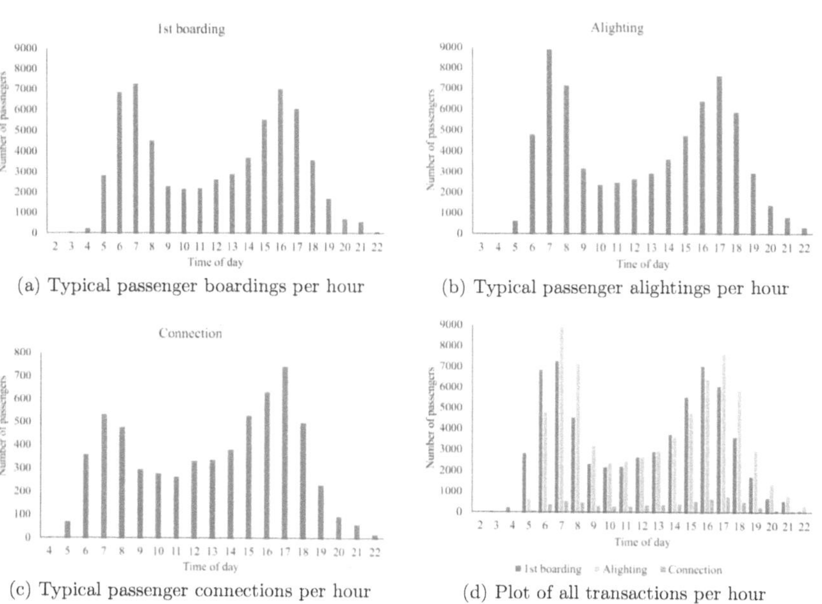

(a) Typical passenger boardings per hour

(b) Typical passenger alightings per hour

(c) Typical passenger connections per hour

(d) Plot of all transactions per hour

Figure 7.1: Plot of a typical daily transactions on the MyCiTi BRT service.

7.2 First application scenario

In this section, the results of some numerical tests conducted with the SBTNOM are presented. The tests involve further evaluating the network solutions in the Pareto front obtained from the model's design. The evaluation of each network solution is done with the Multi-Agent Transport Simulation (MATSim) tool presented in chapter 3. The output of simulating travel demand in MATSim are *events* files. All the actions taken by the agents during the execution of their daily plans on a network is recorded in the file. The events files are then analysed to obtain aggregated values for the performance indicators. The Pareto front can be seen in Figure 7.2.

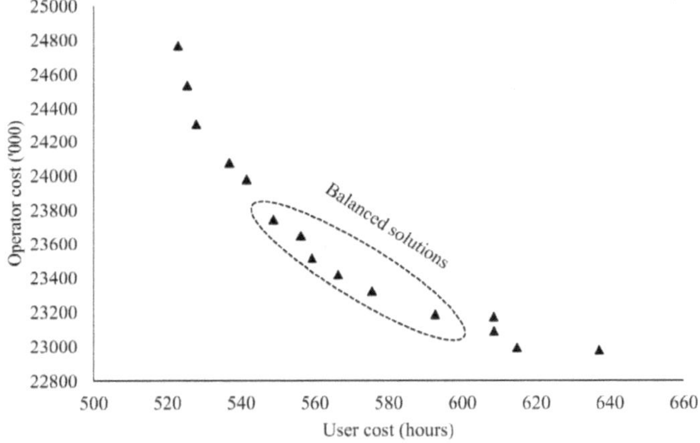

Figure 7.2: Best solutions on the Pareto front after 50 generations iterations.

Visual observation of the plot reveals that the network with the highest travel time also has the least operator cost and vice versa. This is indicative of the *trade-off* between both the users' and operators' perspectives. Users prefer direct trips that reduce their travel time, while operators prefer longer and slightly more circuitous routes that increase the volume of demand they can *potentially* satisfy while reducing their average service costs in the process. The *balanced* network solutions occur within the marked cluster in the middle of the plot in Figure 7.2. In the context of this work, the networks are considered to be balanced as they exhibit the least conflict between the earlier mentioned objectives. In other words, they are the best compromise solutions between the objectives. These solutions are therefore regarded as *efficient* network solutions to the problem. Table 7.1 show the objective function scores for the solutions in the Pareto front.

Table 7.1: Raw objective function values.

Network(-)	User cost(hours)	Operator cost('000)
1	519.79	24767.52
2	520.51	24536.12
3	525.85	24304.72
4	532.02	24073.32
5	535.5	23976.82
6	544.26	23745.42
7	552.67	23648.92
8	555.06	23514.02
9	560.16	23417.52
10	572.36	23321.01
11	587.91	23186.12
12	602.2	23089.61
13	603.01	23172.08
14	609.32	22993.11
15	633.37	22979.07

In the table, network 1 has the least *user cost* or *objective 1* score and will be referred to as the *user-centric* solution. This network also has the highest *operator cost* or *objective 2* score. On the other hand, network 15 has the least operator cost and will be called the *operator-centric* solution. As indicated earlier, the best compromise between the objectives occurs between solution 6 through 11. However, network solution 8 shows the least difference in the normalised values of both objectives. Hence, it may be considered as the best compromise or balanced network solution, since it shows the least conflict between the commuter's and operator's perspectives. To depict this clearly, the above objective scores are normalised to facilitate plotting them on similar scales (see Table 7.2). Subsequently, the normalised scores are ordered and plotted against one another. Figure 7.3 shows a plot of the normalised objective(s) function scores.

Table 7.2: Normalised objective function values.

Network	User cost	Operator cost
1	1.00	10.00
2	1.06	8.84
3	1.48	7.67
4	1.97	6.51
5	2.24	6.02
6	2.94	4.86
7	3.61	4.37
8	3.79	3.69
9	4.20	3.21
10	5.17	2.72
11	6.40	2.04
12	7.53	1.56
13	7.59	1.97
14	8.09	1.07
15	10.00	1.00

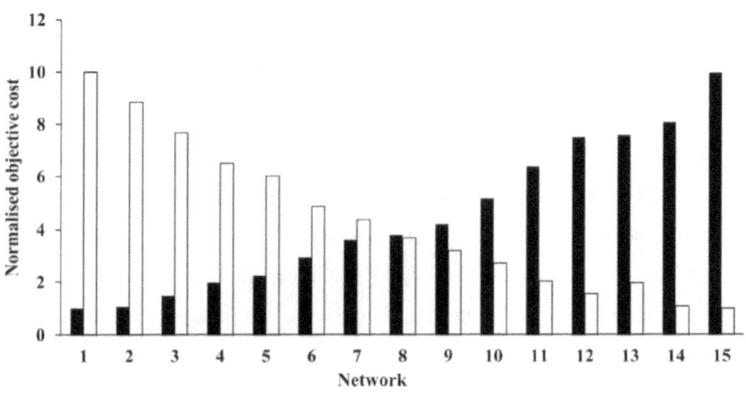

Figure 7.3: Network solutions on the Pareto front showing different compromise solutions.

Having identified these three network solutions — 1 (user-centric), 8 (balanced) and 15 (operator-centric) as proxies for the perspectives mentioned above, they are then isolated for further analysis. The analysis is carried out to measure their performance in terms of different network performance indicators. The indicators used include total satisfied travel demand, total operational cost, network utilisation percentages, unsatisfied demand, vehicle mileage and vehicle hours. The results of the analysis are presented in Table 7.3.

Table 7.3: Aggregate transit network performance indicators for the identified objectives.

Indicators	Network 1 (User-centric)	Network 8 (Balanced)	Network 15 (Operator-centric)
Satis. demand (pax)	34216	31694	29590
Utilisation (%)	88.71	82.17	76.72
Veh. dist (km)	48567.20	45215.15	42452.99
Veh. time (hr)	1618.91	1507.17	1348.43
Op Cost ('000)	2484.14	2316.54	2178.44

In the table, network 1 has the highest satisfied demand and network utilisation; as well as the highest operational cost. This is similar to an optimisation scenario in which the users' objectives are prioritised. In such cases, more passenger demand on direct routes is served. Therefore, circuitous routes and those running through transfer points will be minimal or excluded where necessary. This also means that on average passengers will travel shorter distances to their destination, which will encourage more people to use the service. However, the increase in ridership leads to an attendant increment in the operational frequencies. This maybe, because operators would like to maintain the attractiveness of their service by sustaining a good level of service in terms of travel time, to encourage continued patronage from commuters. Typically, increased operational frequency is known to be a major cost driver for the service operators as the operator will have to deploy more resources (personnel and fleet) on the network.

In contrast, network solution 15 shows an opposite trend to that of the network 1 as it has the least operator cost and least total network utilisation. This is also similar to a case where the operator's objective is prioritised. The results show that the operator has less vehicle mileage and operational hours than the user-centric solution. However, it satisfies less demand. This may be because while transit operators try to maximise network coverage by using circuitous routes, this may ultimately discourage some passenger who wants to use only the direct routes. A network that is skewed in favour of the operator will primarily contain routes that are longer than the preference of users.

Lastly, an optimal transit network solution would contain a mix of direct routes and other more circuitous ones. Hence, the solutions earlier referred to as the best compromise solutions represent a balance between the user and operator perspectives, respectively. Since direct routes reduce the ability of operators to cover demand occurring along more circuitous paths, an optimised solution must compromise between the needs of commuters and service operators. This is revealed in the middle column of Table 7.3 for network 8 where the indicators can be seen to have values between the user and operator perspectives. The results show that the solution is indeed an efficient one as it minimises cost for all stakeholders. The outcomes discussed above are reinforced in Table 7.4 where the balanced network solution has indicator values that compromise between the users' and operators' perspective.

Table 7.4: Average performance indicators at the route level for the identified scenarios.

Indicators	Network 1 (User-centric)	Network 8 (Balanced)	Network 15 (Operator-centric)
No of routes	46	46	46
Route density (pax/route)	743.83	689.00	643.26
Avg op cost ('000)	54.00	50.36	47.36
Avg veh time (hr/route)	35.19	32.76	29.31
Avg veh dist (km/route)	1055.81	982.94	922.89

This result shows that the balanced solution is the most attractive for all stakeholders, and it also offers better access to public transit services. Ultimately, depending on the what a policymaker intends to achieve they can easily apply decision support tools such as a Multi Criteria Decision Analysis (MCDA) to the obtained results, to arrive at other trade-off solutions from the set of non-dominated solutions to match their priority. These results show that Simulation-based Optimisation (SBO) can indeed yield reasonable network solutions when used in the Transit Network Design (TND) process. They also show that Agent-based Simulation (ABS) can play an essential role in the process. In conclusion, the potential of the Simulation-based Transit Network Design Model (SBTNDM) to use big data and other technological advances unfolding in the transit sector today make it viable for modelling large-scale and complex transport scenarios. The images in Figures 7.5 to 7.7 show the three route networks referenced in the discussion above. In Figure 7.8 all three networks are overlaid against one another. They are extracted from the highlighted part of the base network shown in Figure 7.4.

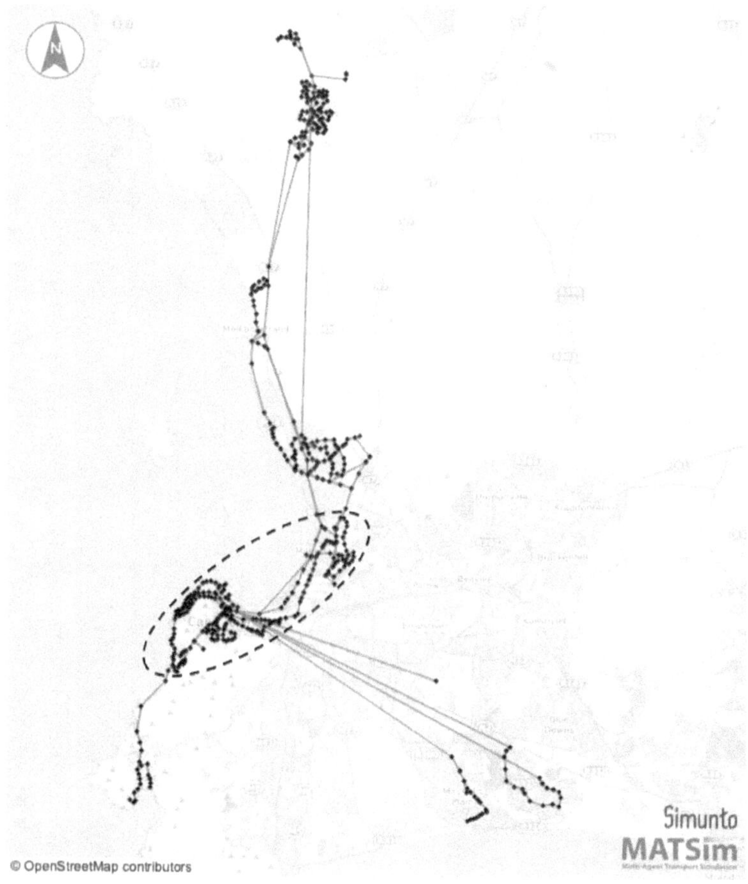

Figure 7.4: Map of the MyCiTi BRT base network.

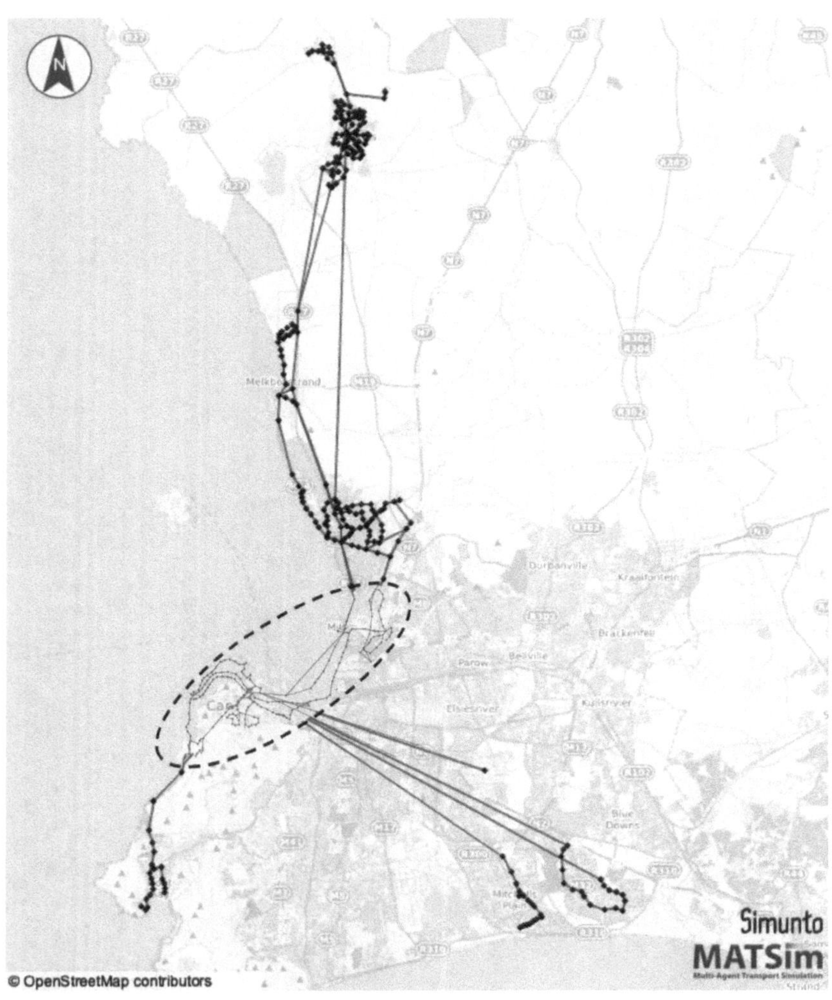

Figure 7.5: Depiction of the network that favours the commuter (Network 1).

Figure 7.6: Depiction of the network that favours the operator (Network 15).

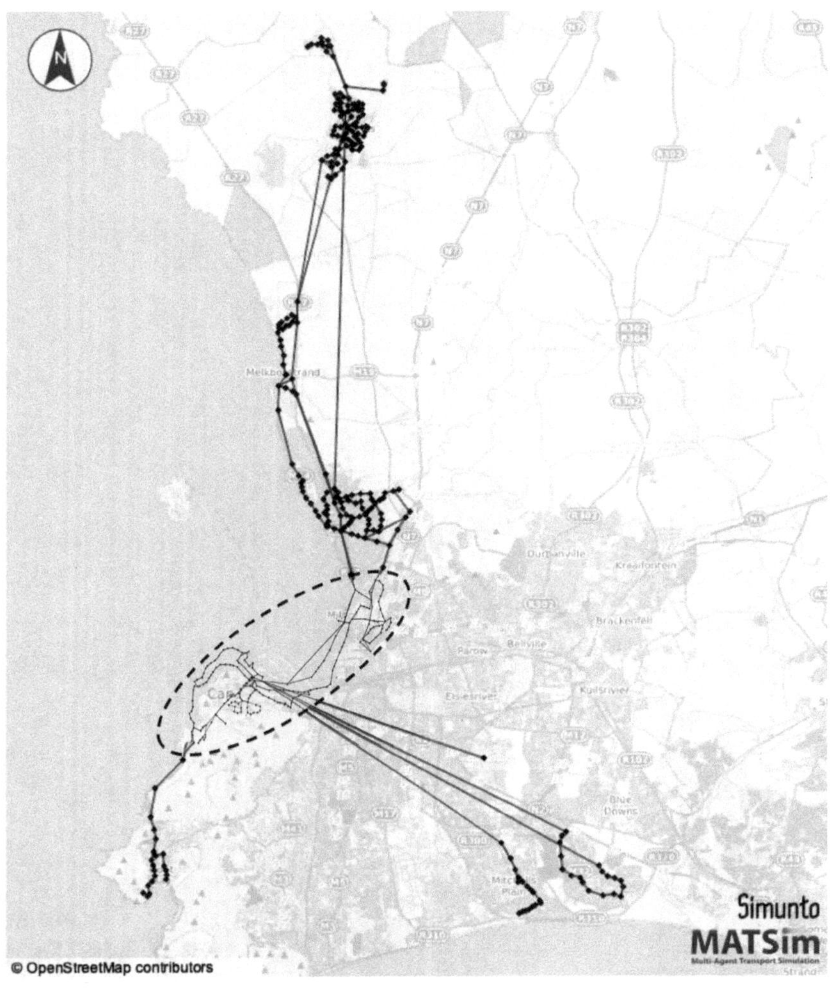

Figure 7.7: Depiction of the balanced network (Network 8).

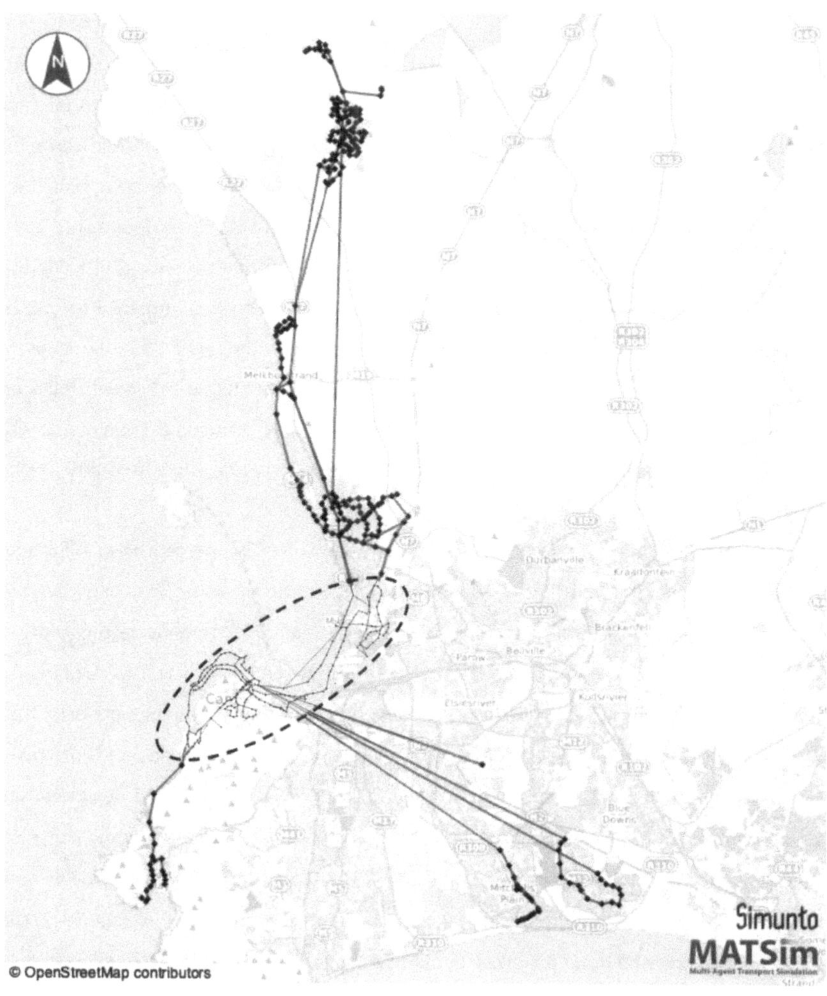

Figure 7.8: An overlay of the three networks.

7.3   Second application scenario

This case study involves designing a public transport network that satisfies the revealed passenger ridership on the MyCiTi service over one year. The objective of this section is to demonstrate the ability of SBTNDM to design a good transit network under different demand conditions that occur within an operational year of the MyCiTi BRT service. To do this, AFC passenger travel demand data for a 12 month period between September 2015 and August 2016 has been collected. The data is analysed to obtain the 85th percentile of total travel demand in each month. Subsequently, the demand data for each of the 12 months was used to design transit network solutions. The designed networks may then be compared to obtain the most efficient network solution for the stated calendar year. The concept of 85th percentile is used mostly in traffic engineering for setting travel speed limits on highways and other travel time reliability studies. In that context, it is known as the speed at which 85 percent of all drivers will travel at or below under free-flow traffic conditions.

Similarly, given a distribution of travel demand, the 85th percentile or $Q85$ may be considered as all the values of travel demand that are up to or less than the 85% of the values in the distribution. Using the 85th percentile may be a better solution than the maximum demand scenario because, in the context of the MyCiTi BRT network, the maximum demand may not always be obtainable on the network. It is a widely observed trend in public transportation that there is less demand for travel on weekend and holidays in comparison to working days. To this end designing for maximum or minimum demand scenarios will not be ideal. However, while the 85th percentile demand is designed for here, it should be noted that as a very flexible tool the SBTNDM can be used to design for various demand scenarios. Consequently, the same analysis can be done for other demand scenarios such as the 90th 95th or 99th percentiles respectively. Policymakers will ultimately decide for a demand scenario in consultation with other key stakeholders that use the network. Table 7.5 shows the total 85th percentile daily ridership for the 12 months under consideration.

Figure 7.9a to 7.10h shows a plot of the passengers transactions (boarding,

Table 7.5: 85th percentile daily ridership for 12 months.

Year	Month	Ridership (pax)
2015	Sept	36890
2015	Oct	37094
2015	Dec	37431
2015	Nov	38116
2016	Jul	39540
2016	Jan	40325
2016	Jun	40866
2016	Aug	41414
2016	Feb	42933
2016	May	42995
2016	Apr	43260
2016	Mar	43746

alighting and connections) for the 85th percentile data captured in the table.

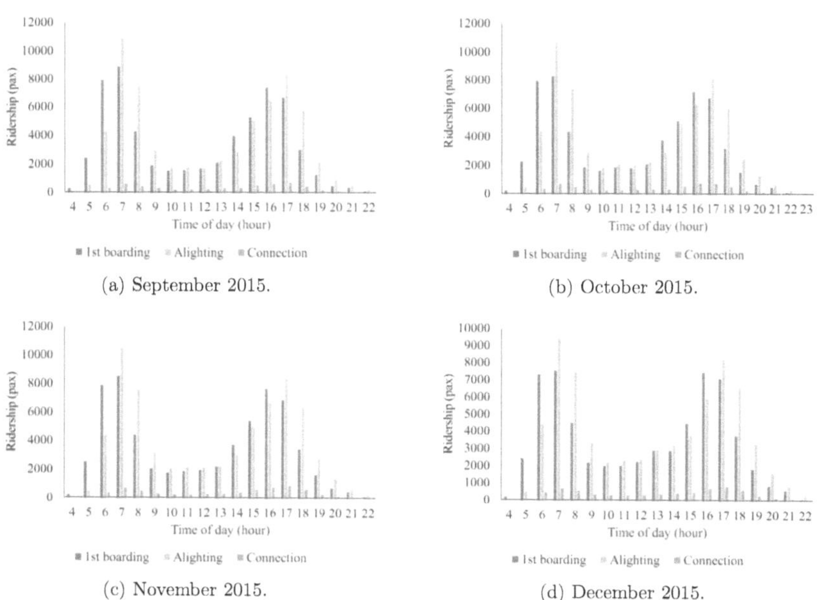

(a) September 2015.

(b) October 2015.

(c) November 2015.

(d) December 2015.

Figure 7.9: 85th percentile daily passenger ridership on the MyCiTi BRT.

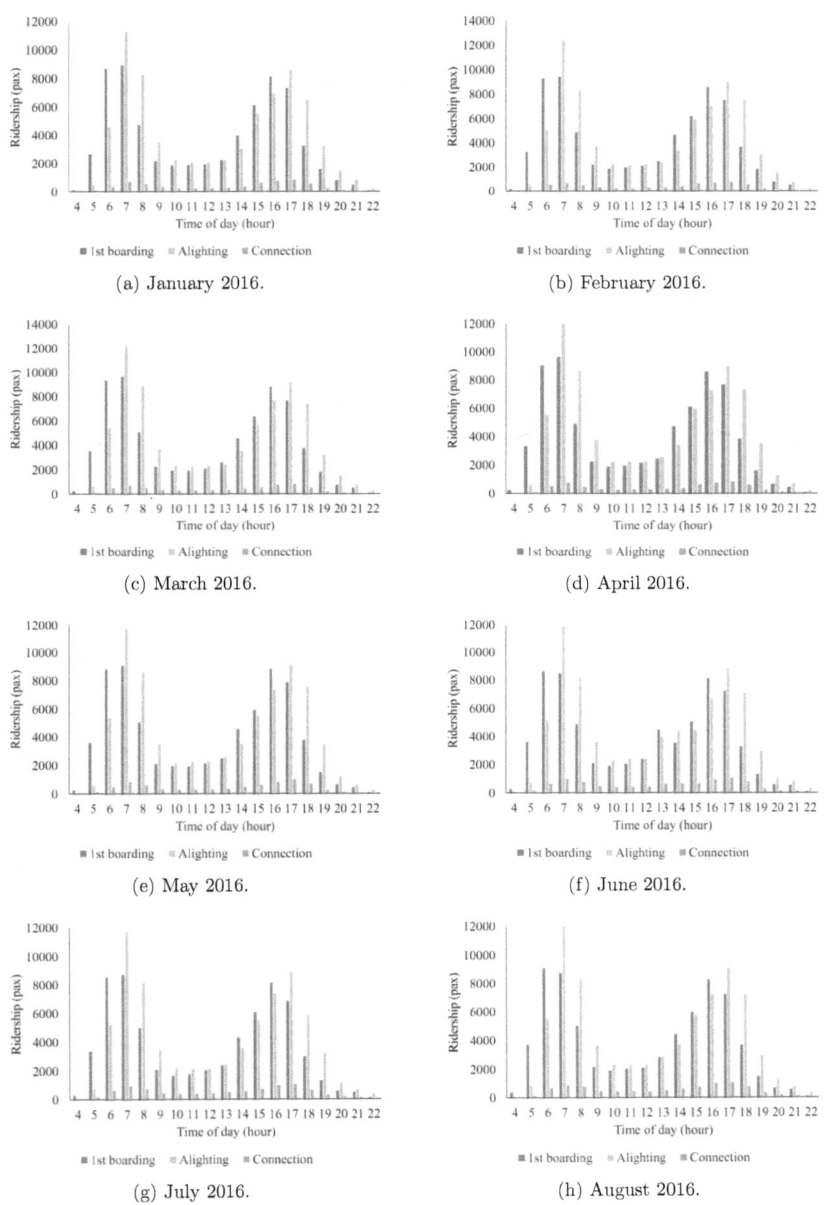

(a) January 2016.

(b) February 2016.

(c) March 2016.

(d) April 2016.

(e) May 2016.

(f) June 2016.

(g) July 2016.

(h) August 2016.

Figure 7.10: 85th percentile daily passenger ridership on the MyCiTi BRT cont'.

Here, the SBTNOM is used to perform 12 network design exercises using the data collected for each month. Subsequently, the results are analysed in a way that is similar to that discussed in the first test case is. Table 7.6 shows the least objective scores for the user and operator objective only as they occur in the respective design for each month within the design period. The best solution for the user perspective occurs in September 2015, while that for the operator perspectives occurs in March 2016.

Table 7.6: The non-dominated solutions with the least objective scores for the user and operator for the 12 month period.

Date	User cost(pax)	Operator cost ('000)
Sept 2015	497.24	23904.67
Oct 2015	503.95	23931.88
Nov 2015	520.49	23822.21
Dec 2015	515.77	24028.28
Jan 2016	576.02	24014.78
Feb 2016	616.04	24028.39
Mar 2016	635.09	21013.89
Apr 2016	614.52	23604.41
May 2016	623.47	24000.74
June 2016	615.15	24000.74
July 2016	565.74	23797.42
Aug 2016	593.62	23700.92

Recall that the user-centric and operator-centric solutions are the two extremes of the Pareto front. They are referred to as such because they both perform best in terms of the user and operator goals respectively. Usually, in the Pareto front of a typical Transit Network Design Problem (TNDP) and in the broader case of other Multi-objective Optimisation Problems (MOPs), there are other trade-off solutions between these extreme points which are all considered as optimal. This implies that any of these solutions corresponds to a degree of conflict or trade-off between the different objectives. However, the availability of many trade-off solutions does not alleviate the need for a policy maker to find a single solution that best meets their intended requirement. Therefore, more commonly, it is the case that Decision Makers (DMs) require a deeper understanding of the problem and the available solutions to determine the best one. Deb (2001) refers to this as *higher-level information*, which often involves the consideration of other secondary factors and attributes of the problem and the non-dominated set of solutions.

In terms of obtaining the best compromise network from the set of 12 networks, Figure 7.11a to Figure 7.12h shows results obtained from optimising the networks. An observation of the results shows that 4 networks show the possible least compromise, namely networks 5 and 6 from October 2015 (see Figure 7.11b) and networks 6 and 7 for March 2016 (see Figure 7.12c). Therefore, in line with the preceding discourse, any choice of a final trade-off solution would depend on what a policymaker intends to achieve. They can easily apply decision support tools such as a MCDA to the obtained results to identify the best one given their priority.

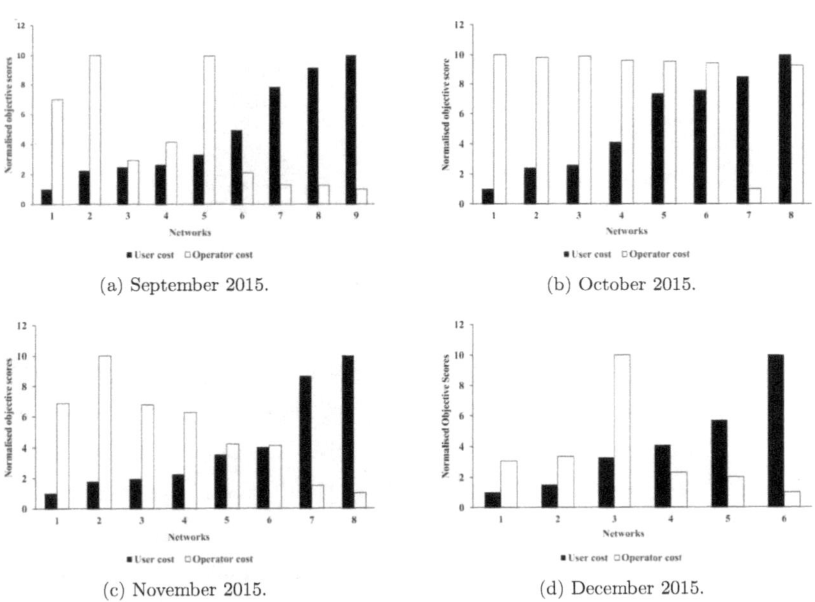

(a) September 2015.

(b) October 2015.

(c) November 2015.

(d) December 2015.

Figure 7.11: Network results for the 12 months.

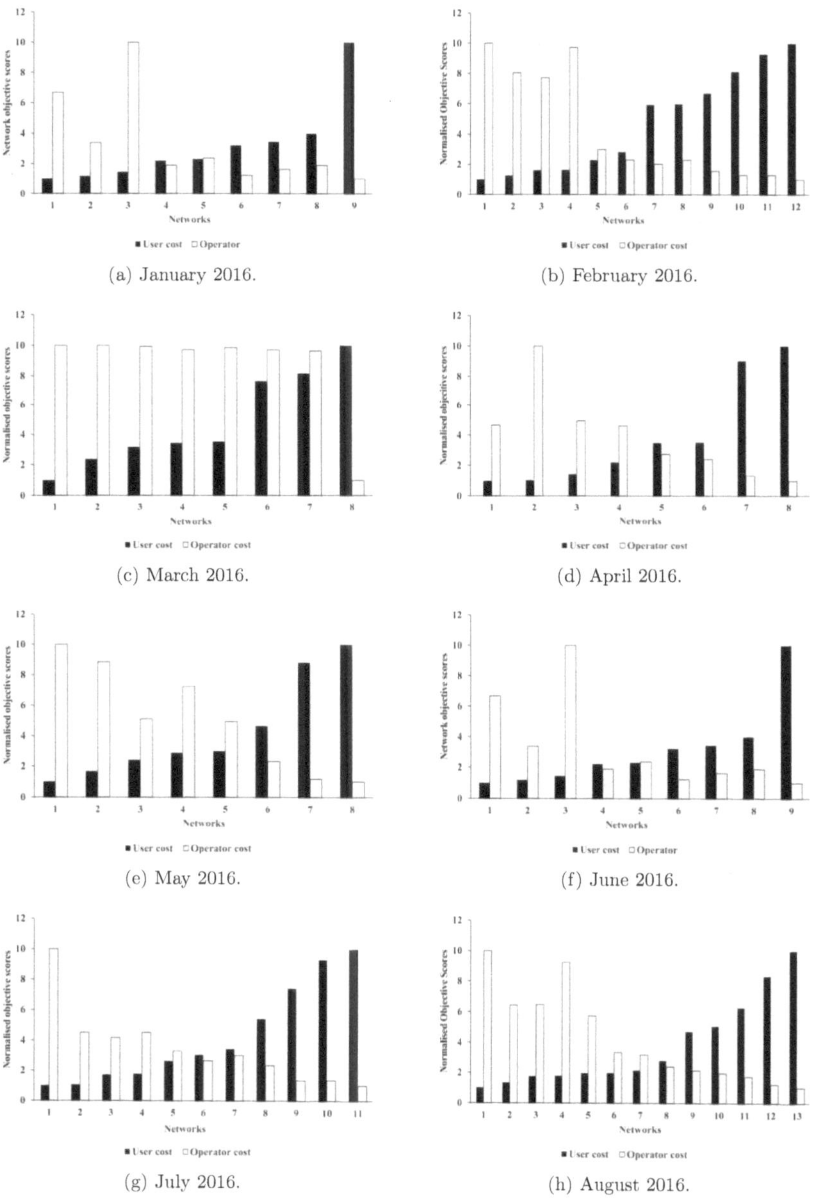

(a) January 2016.   (b) February 2016.

(c) March 2016.   (d) April 2016.

(e) May 2016.   (f) June 2016.

(g) July 2016.   (h) August 2016.

Figure 7.12: Network results for the 12 months cont'.

These results show that SBTNDM is capable of designing robust networks which can address the variability of travel demand over the one year design period under review. This implies that the model can suitably be used to design public transportation networks for travel demand data spanning a longer time frame. Hence, the model can find application in strategic or long term public transport network planning and design in Cape Town. From decision support and policy making standpoints, the SBTNDM can potentially serve as a tool in the tool-kit of the TDA in Cape Town. This is because the model can identify many viable transit network alternatives and show their trade-off for various network stakeholders. Therefore, this will allow transportation experts and authorities to choose one of the alternatives for further consideration in terms of its accessibility, economic or environmental viability.

In conclusion, the numerical tests carried out in this chapter, show that Agent-based Models (ABMs) can yield good results when used in the transit network design process. Also, the potential of the SBTNOM to model complex network scenarios with big data and other technological advances unfolding in the transit sector today sets it apart. This is because many of the static models used today are not capable of incorporating newer technological advances in their network design methods. This makes the SBTNOM a tool for the future in terms of transportation planning and engineering endeavours.

7.4 Summary

In this chapter, numerical tests were conducted on the SBTNOM. The objective was to demonstrate the model's ability to design large-scale and realistic public transportation networks. Two tests were conducted to demonstrate this. The first test analysed three trade-off solutions obtained from the Pareto front, of a network design exercise conducted with the SBTNOM. The three solutions represented a user-oriented viewpoint, the operator viewpoint and a more balanced view. In the second test, a more strategic network design approach was taken, which involved designing a robust network on the MyCiTi BRT with ridership data collected over a year. The results of these tests demonstrate that the model is indeed capable of designing public transit networks. This highlights the model's potential importance to the transit network design and planning endeavours in Cape Town.

Chapter 8

Summary and conclusion

8.1   Summary

Public transportation networks facilitate human mobility; they also play a critical role in the smooth exchange of goods, services and information. Indeed, the growth and prosperity of any society can easily be linked to the state of its transportation networks. This is because the non-existence of a network or the existence of an inefficient one can decrease people's mobility and their ability to access opportunities like jobs, healthcare, education and recreation. Furthermore, public transportation networks are drivers of growth, since they facilitate the creation of jobs and enhance the economic outlook of cities by attracting significant financial investment from government and other investors. These benefits may, therefore, be lost if attention is not paid to transit network improvement and upgrades when it is needed.

The current public transportation operational environment in South Africa reveals inefficiencies that have limited the country's ability to maximise some of the earlier mentioned benefits. The status quo has resulted in low passenger ridership in public transport and a high dependence on private cars which has worsened congestion and environmental pollution. Furthermore, due to the difficulties involved in balancing operational service levels with cost, many public transport systems in South Africa suffer from low productivity, high costs, and a need for large government subsidies. Therefore, to address some of these issues, public transportation networks must be given priority in terms of improving and upgrading them.

The main objectives of this work have been to develop a Transit Network Design (TND) approach that improves on the existing methods in the literature and to apply the model to a Transit Network Design Problem (TNDP) case study in the city of Cape Town, South Africa. The central questions of the research were *How can a Simulation-based Optimisation (SBO) solution model for the transit network design problem be developed through integrating multi-objective optimisation and simulation models?* and *How can the optimisation model be used to design public transit networks in the city of Cape Town, South Africa?*

These questions have been addressed in the **book** by developing a Simulation-based Transit Network Optimisation Model (SBTNOM), that combines the Agent-based Model (ABM) and Multi-objective Optimisation (MOO) in the design of optimised public transit networks. Subsequently the model has been tested computationally and then applied to the optimised design of the MyCiTi Bus Rapid Transit (BRT) network in Cape Town. The tests and application, shows the model's suitability for the design of large-scale public transit networks.

The Simulation-based Transit Network Design Model (SBTNDM) combines Multi-Agent Transport Simulation (MATSim) and Non-dominated Sorting Genetic Algorithm (NSGA-II). The former evaluates a network solution by simulating the individual travel experience of passengers on the network, while the latter performs the actual network optimisation based on the results of the simulation. Another novelty in the work has been to define an encoding scheme in the network optimisation model. This encoding scheme allows for the full representation of a transit network and its schedules within the optimisation framework, which then facilitates the simultaneous optimisation of the network and its schedules.

Finally, the so-called SBTNDM has undergone rigorous computational testing. It has also been applied to the case study of the MyCiTi BRT network in Cape Town, South Africa. The results of the tests and application reveal that the model is capable of designing good compromise network solutions that are robust and reflect the objectives of the different network stakeholders like passengers and operators.

8.2 Limitations of the research

There is room for improving the SBTNDM as it only models interactions that are internal to the transit network without considering external factors that can influence what happens on the system. These include the activities of passengers that give rise to their demand for the service and the connecting trips made by passengers to get unto the BRT system. However, in the context of this **book**, this is plausible because the scope of application is to a BRT network, which has dedicated infrastructure alignment that often operates with little or no interactions with the broader transportation system. The other limitation in this work is that the data used in this research is the revealed travel demand of the MyCiTi BRT. This implies that the optimal network solutions obtained will only reflect the current demand on the BRT system and not capture the future growth in travel demand on the system. It is noteworthy that the data collection and sampling required to create a future demand scenario input is financially expensive and thus, outside the scope of this **book**. Furthermore, the SBTNDM is a versatile framework that can produce different network solutions based on the input demand. To this end, the travel demand input of the model can be replaced, and optimised network solutions will be obtained based on the new data.

8.3 Future research possibilities

In terms of the possible research directions that may extend from this work in the future, the primary consideration is to widen the application of the SBTNDM to a hybrid network context, to improve modal integration in Cape Town. In the city, different public transport services have different ownership structures and operational procedures. Consequently, the various services plan their respective transit routes and operational schedules without recourse for other modes. This results in a mixed and almost self-organising operating environment known in Salazar Ferro et al. (2013) as a *hybrid* network. Better coordination between the different modes will potentially be achieved by integrating their services. This, however, raises several questions, including how such an integrated network may be designed.

Therefore, after successfully applying the SBTNDM to a unimodal transit network design context the model may then be extended to this hybrid network context where the imperative will be to design public transport networks that address the issue of inter-modal coordination in Cape Town. This will then inform the further development of a fully integrated public transport system involving all common public transport modes in the city. The major challenge in this endeavour lies in creating the inter-modal passenger trip chains and schedules, which includes the BRT and other modes. Cognisance is given to the fact that only the BRT service in Cape Town uses an Automated Fare Collection (AFC) system. However, it is hoped that the required data can be obtained through surveys and liaison with relevant government and private agencies in the city.

Another possible direction for expanding the SBTNDM in the future, maybe in its use for accessibility-based transit network optimisation. Historic and conventional transportation planning policies in Cape Town have focused on mobility-centred planning, which primarily evaluates a transportation system on its ability to facilitate the efficient movement of people and goods. Therefore, this philosophy considers as beneficial, increases in vehicular travel volumes and speeds. However, past experiences in Cape Town show that this type of planning favours automobile-oriented policies and land-use patterns which increases the potential for transport related externalities like accidents and environmental pollution. In more recent times, there has been a shift towards accessibility-based planning, which gives more significant consideration to the ease with which people can reach their desired activity location or opportunity. However, regardless of the transportation policy in place at any time, one point of interest for the policymaker is to have a tool that enables them to design their network and test different assumptions on the model. Hence, the SBTNOM with its agent-based simulation approach would be suited to describe the dynamics of accessibility modelling correctly. The outcome (travel demand) of modelling people's accessibility can then be used as an input for the optimisation of the transport network using the SBTNOM.

8.4   Conclusion and recommendations

The SBTNDM developed in this book is a versatile tool which offers a new approach to transit network design. It adopts new and advanced technology-driven tools for addressing some of the challenges associated with planning, designing and operating efficient public transportation networks. This is important because as technology causes *disruptions* in various facets of society today, it is crucial to adapt to these changes.

Instances of these disruptions in the broader transport sector is the advent of concepts like ride-sharing, autonomous vehicles and Mobility As A Service (MAAS). These have all been developed on the back of technologies like Global Positioning System (GPS), wireless communications and sophisticated geographic sensing. Also, the coming of big data which is collected from AFC systems, among others is changing how transport-related data is collected, analysed and used for policy implementation. Also, high-speed portable computer processing and Artificial Intelligience (AI) now makes it possible to solve more complex transportation-related problems with more incisive data leading to better results. Collectively, these technologies are influencing different aspects of transportation and making changes that may not have been predicted a short while ago.

Looking specifically at public transportation planning, it is not hard to see some of the changes that have arisen as a result of the adoption of technology in many spheres of our lives as humans. For instance, Polzin (2016) outlines some changes technology has made and how it is affecting the transportation modelling process. In terms of trip generation, the advent of video conferencing, for example, means that some interactions, that would have compulsorily given rise to a trip in the past can now be done virtually without travelling. Likewise, in trip distribution the choice of a person's destination is now assisted by the ability of a traveller to obtain better information about that destination; a traveller can now determine the availability and price of a commodity or service before travelling to a retail location. Relating to mode choice, people obviously now have more modes of transport at their disposal today, implying that public transport now competes with more modes than the traditional *automobile*. Lastly, trip assignment can now be done with GPS in on-board navigation systems or digital mapping, and with real-time information like General Transit Feed Specification (GTFS). Therefore, to remain effective and relevant, the classic transport modelling process will have to adapt to these changes, even though the fundamental principles of the model remain the same.

The SBTNDM therefore adopts technology-driven tools like agent-based modelling, big data and AI to develop network solutions that are robust and adapt well to the changing landscape of travel behaviour. As a decision support tool, the model will be useful in guiding policymakers in Cape Town in making policy decisions that are relevant to the transportation context and realities of today. In light of this, the author recommends that the model be adopted as a public transport network planning and design tool in the City of Cape Town (CoCT).